Introduction to Environmental Modelling

Jo Smith and Pete Smith

OXFORD

UNIVERSITY PRESS

OXFORD

UNIVERSITY PRESS

Great Clarendon Street, Oxford OX2 6DP

Oxford University Press is a department of the University of Oxford.
It furthers the University's objective of excellence in research, scholarship,
and education by publishing worldwide in

Oxford New York

Auckland Cape Town Dar es Salaam Hong Kong Karachi
Kuala Lumpur Madrid Melbourne Mexico City Nairobi
New Delhi Shanghai Taipei Toronto

With offices in

Argentina Austria Brazil Chile Czech Republic France Greece
Guatemala Hungary Italy Japan Poland Portugal Singapore
South Korea Switzerland Thailand Turkey Ukraine Vietnam

Oxford is a registered trade mark of Oxford University Press
in the UK and in certain other countries

Published in the United States
by Oxford University Press Inc., New York

British Library Cataloguing in Publication Data
Data available

Library of Congress Cataloging in Publication Data
Data available

Typeset by Laserwords Private Limited, Chennai, India
Printed in Great Britain
on acid-free paper by
Antony Rowe Ltd, Chippenham, Wiltshire

ISBN 978–0–19–927206–8

10 9 8 7 6 5 4 3 2 1

■ CONTENTS

PREFACE   viii

ACKNOWLEDGEMENTS   x

1   Introduction   **1**

1.1   What if there was life on Mars?   1

1.2   What is a model?   2

1.3   Why use models?   4

1.4   Which model should I use?   8

   1.4.1   Determining what type of model to use   9

   1.4.2   Determining what type of mathematics to use in the model   15

1.5   Choosing an existing model   20

1.6   How is a model made?   21

   Summary   23

   Problems   24

2   How to develop a model   **26**

2.1   Choose the type of model   27

   2.1.1   Why am I doing this?   27

   2.1.2   How should I do this?   27

   2.1.3   The principle of parsimony   29

   2.1.4   What do I do now?   30

2.2   Draw up a conceptual model   31

   2.2.1   Draw a picture   32

   2.2.2   List all hypotheses   33

   2.2.3   List all assumptions   33

   2.2.4   Set the boundary conditions   33

2.3   Attach a mathematical model   35

   2.3.1   The parts of a mathematical model   36

   2.3.2   Linking together fixed parameters and input variables   36

   2.3.3   Choosing the mathematical approach to derive the model   38

   2.3.4   Example   40

2.4   Construct a computer model   45

   2.4.1   General spreadsheets   46

   2.4.2   Specialist modelling software   52

   2.4.3   High-level programming languages   55

2.5   And then . . . ?   61

   Summary   62

   Problems   63

   Further reading   65

References 67
Web links 67

3 **How to evaluate a model** **69**

3.1 Decide what type of evaluation is needed 70

3.2 Plot the results (graphical analysis) 71
  3.2.1 Plots to reveal the accuracy of the simulation 72
  3.2.2 Plots to illustrate the behaviour of the model components 76
  3.2.3 Plots to establish the important model components 78

3.3 Calculate the accuracy of the simulation (quantitative analysis) 81
  3.3.1 Analysis of coincidence 84
  3.3.2 Analysis of association 94
  3.3.3 An example of the use of statistics to assess the accuracy of a model: the model of soil
    carbon change on Mars 98

3.4 Examine the behaviour of the model (sensitivity analysis) 106
  3.4.1 What is sensitivity analysis and why is it important? 106
  3.4.2 Methods used in sensitivity analysis 107
  3.4.3 Expressing sensitivity 107

3.5 Determine the importance of the model components (uncertainty analysis) 109
  3.5.1 What is uncertainty analysis and why is it important? 109
  3.5.2 Methods used in uncertainty analysis 110
  3.5.3 Representing variation in the input parameters and model outputs 110
  3.5.4 Expressing uncertainty 111

3.6 And then . . . ? 113

Summary 114

Problems 115

Further reading 117

References 117

Web links 118

4 **How to apply a model** **119**

4.1 Scientific representation 122
  4.1.1 Who is the end-user? 123
  4.1.2 How is the model used? 124
  4.1.3 Guard against input error 126
  4.1.4 Guard against misinterpretation of the results 126
  4.1.5 Documentation 127

4.2 Expert and decision support systems 128
  4.2.1 Who is the end-user? 129
  4.2.2 How is the model used? 130
  4.2.3 Guard against input error 130
  4.2.4 Guard against misinterpretation of the results 132
  4.2.5 Documentation 132

4.3 Risk assessment 134
  4.3.1 Who is the end-user? 135
  4.3.2 How is the model used? 136

	4.3.3 Guard against input error	137
	4.3.4 Guard against misinterpretation of the results	137
	4.3.5 Documentation	138
4.4	Spatially-explicit applications	139
	4.4.1 Who is the end-user?	139
	4.4.2 How is the model used?	140
	4.4.3 Guard against input error	143
	4.4.4 Guard against misinterpretation of the results	144
	4.4.5 Documentation	144
4.5	Epilogue	146
	4.5.1 How has life on Mars been improved?	146
	4.5.2 The real-Earth applications of the models used in this book	147
4.6	So is this the end?	150
	Summary	152
	Problems	156
	Further reading	157
	References	157
	Web links	158

Appendix 1 Solutions to problems

1.1	Solutions to problems in Chapter 1	159
1.2	Solutions to problems in Chapter 2	160
1.3	Solutions to problems in Chapter 3	167
1.4	Solutions to problems in Chapter 4	172

Glossary	176
INDEX	179

■ PREFACE

About this book

This book is about the philosophy, methods and issues surrounding environmental and ecological modelling. Environmental and ecological systems are complex, and modelling often provides the only means by which a researcher can make sense of results and extrapolate them to other environments. Modelling allows research findings to be extended to the real world, and promises to help solve many problems facing us in the twenty-first century. Through the development of model-based expert systems, decision support systems, risk assessment methods and regional modelling techniques, non-experts ranging from farmers to policy makers are already using models every day to solve environmental and ecological problems. To be able to apply environmental and ecological science in the real world, we need to understand how models are conceived, developed and tested.

This book provides you with the skills needed to develop your own models, to understand how environmental and ecological models should be used, and to decide how far they should be believed. The book is aimed at the final year in the undergraduate curriculum, and new researchers who need to learn about modelling. It is accessible to a novice in modelling, and equips the reader with all the essential knowledge for the practical application of modelling. Ability in mathematics to GCSE, Scottish Standard Grade or equivalent is assumed.

This book does not focus on the theoretical niceties of modelling, but tells and shows you *how to* model. It explains the theory of modelling from first principles to a level that is sufficiently advanced to allow its application in research. You will gain a thorough understanding of modelling theory, and, through worked examples, will acquire the practical ability to develop new models and test your own applications. We describe methods used in model development, techniques used to evaluate the behaviour of models, and issues surrounding their application.

Bringing modelling to life: worked examples

The techniques are illustrated throughout by **worked examples** from an imaginary world on Mars. In this world, examples are simple enough to allow the principles to be conveyed without the need to go into too much background! This will give you a full theoretical understanding as well as practical experience in modelling. The book concludes with examples of model application in the real world, drawing out the modelling principles already described in the book.

Other learning features

The book also includes three other learning features, in addition to the worked examples mentioned above.

- **Key points** are highlighted throughout the chapters.
- **Self-check questions** appear at the ends of many sections, giving the reader the opportunity to check their understanding of the material covered in the preceding section.
- **Problems** appear at the end of each chapter. The problems build on the self-check questions, and encourage the reader to draw on their understanding of the various topics covered in the preceding chapter.

Online Resource Centre

This book is accompanied by an Online Resource Centre at

www.oxfordtextbooks.co.uk/orc/smith_smith/

The Online Resource Centre features additional worked examples to complement those in the text. It is essential to study these worked examples if the book is to provide practical experience as well as theoretical understanding.

The Online Resource Centre also features the following.

For registered adopters of the book:

- figures from the book available to download.

For students:

- a library of web links cited in the book;
- hyperlinks to the full text of primary literature articles cited in the book, where available.

■ ACKNOWLEDGEMENTS

We are very grateful for the essential contributions to this book from our colleagues who are expert in a wide range of different modelling techniques. These inputs have been through direct contributions to the text and website, as well as through the more diffuse inputs of ideas and concepts over the years. Thank you to Peter Lefelaar (Wageningen University, the Netherlands), Gianni Bellocchi (formerly Agriculture Research Council, Bologna, Italy), Matt Aitkenhead (formerly Macaulay Institute, UK), and John Durban, Paul Thomson, John Townend, Toby Marthews, Martin Wattenbach, Pia Gottschalk and Mike Haft (University of Aberdeen, UK). Thank you also to Tom Addiscott and David Jenkinson (Rothamsted Research, UK)—former bosses, mentors and great friends!

Figure acknowledgements

We gratefully acknowledge the help of Nick Crowe in drawing the cartoons that appear throughout the book.

<table>
<tr><td>

1
</td><td>

Introduction
</td></tr>
</table>

1.1 What if there was life on Mars?

Mars is a dead planet. All of the information we have from remotely sensed data and, more recently, from landing craft, show that, even if life did ever exist on Mars, it does not now. But was not the Mars of 1960s B-movies more intriguing? Imagine a Mars that is very much like Earth, teeming with life and facing similar environmental and ecological problems to those we face here. The setting for this introduction to environmental and

The 'real' Mars

Mars is NOT a dead planet! World governments do not want us to know and cover up the truth using staged pictures of a dead planet.

Humans have been studying Mars from a distance since the Roswell Incident in 1947 when a Martian spaceship crashed in the New Mexico desert near the town of Roswell.

Some of the data used in this book come from the alien spacecraft,...

...other data come from landing parties,...

...and some are remotely sensed by a series of robot probes on Mars.

ecological modelling is such a Mars. As you use the book and solve the problems, the story of the environmental and ecological problems faced on Mars will unfold, given in a storyline that appears through the text as picture boxes. The picture boxes not only provide the examples for which models will be developed, tested and applied, but they also act as markers to help you skip through the book to find specific subjects in the main text.

Through the picture boxes, we will present a number of environmental and ecological problems faced on this imaginary version of Mars, which could be addressed using models. You will be given the results of previous modelling attempts and you will discover why those models failed. You will then be given the opportunity to develop new approaches to do a better job. On the way, you will learn how to model and how to apply good modelling practice. The book uses guided problem solving to develop modelling skills. It will allow you to learn the essential philosophies and techniques of environmental and ecological modelling in a scientifically rigorous way by learning and applying these techniques to solve problems on Mars. All of the skills you learn from using this book will be applicable to real-world problems—indeed these techniques were developed on Earth!

Mars needs YOU!

Previous attempts to model the environment on Mars have failed. Your job would be to lead the team of exobiologists (people who study life on other planets) to develop models to help address the environmental and ecological problems on Mars. Good luck at the interview!

1.2 **What is a model?**

A model is a simplified representation of reality; this definition applies equally to a physical model, such as might be sculpted in clay, and a mathematical model, constructed using mathematical equations (Fig. 1.1).

When creating a model, a sculptor will observe reality, decide which aspects of reality are important, and use a selection of techniques to represent these important aspects; this representation is the sculpted model. A mathematical modeller follows exactly the same procedure; the only difference between the mathematical modeller and the sculptor is in the methods they use. Like the sculptor, a mathematical modeller observes reality, but this is usually done by some form of experimentation, not just by eye. The mathematical modeller decides which aspects of reality are important to the representation. Section 1.4 introduces a systematic procedure for selecting the aspects that are important, in which the research question is translated into a hypothesis, which then guides the modeller as to the type of model needed, and the important characteristics of the model. As

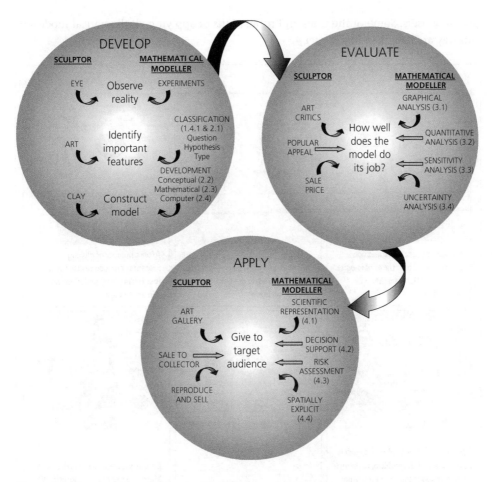

Figure 1.1 How to make a model: a simplified representation of reality.

will be discussed in Chapter 2, the mathematical modeller then uses a combination of mathematical techniques and computer programming to represent these important aspects of reality as equations; this representation is then the mathematical model.

The value of a model lies in its ability to do what it was created to do. Critics may evaluate a classical sculpture according to how much it looks like the subject, whereas an impressionist sculpture is evaluated on an entirely different basis. Similarly, a mathematical model should be evaluated with respect to its ability to achieve its objectives. If it is a quantitative model, the evaluation should also be quantitative. Different aspects of model evaluation are discussed further in Chapter 3.

The value of the model can only be fully realised if it is made accessible to its target audience. A sculptor might do this by exhibiting the model in an art gallery, selling it to a collector, or making reproductions to be more widely sold. Due to the ease of exact reproduction provided by computers, mathematical models are usually packaged in some form that makes them easy to use, and to be reproduced and distributed to potential users. Some of the trials and tribulations of applying mathematical models in this way are described in Chapter 4.

1.3 Why use models?

Many people regard modelling as a combination of difficult mathematics and technically demanding computing. People who spend too much time with models are often classified by others as '**modellers**' and are considered to be rather strange. However, modelling is a ubiquitous tool that is widely used in all branches of science, although it is often known by a different name. One of the most elegant models in history (described as a 'table') was developed by Mendeleyev to describe the electronic structure of elements. The model is, of course, the periodic table, and was used to predict the properties of many elements before they had even been discovered. Without this model, the discovery of many elements would have been delayed for many more years.

In fact, modelling is something that we all do every day. Take the example of choosing a queue at the railway station. We look at the length of the queue, and assess how quickly each queue is moving; in the blink of an eye, we have constructed a model of the time it will take to get to the counter, entered the input data (the relative speed at which the two queues are moving), and obtained the result upon which we choose the best queue to join. In doing this, we have mastered the main stages of model conception and development. These instinctive thought processes are disentangled when we look at model development in Chapter 2.

Many of us even go on to evaluate the model, by noticing which person is at the back of a neighbouring queue, and checking that we do in fact reach the front before them! This evaluation process is rationalised in Chapter 3. Modelling is something that comes naturally to us all, but this is precisely the reason why it is so often misused and misrepresented. Issues surrounding application (and misapplication) of models are discussed throughout this book and are the focus of Chapter 4.

If we are to make proper use of our inherent capability to model, we must have an unambiguous idea of what the model is intended to do (i.e., the **scope** of the model), follow procedures for model development, evaluation and application, and understand that the natural world is unpredictable. After all, who could have predicted that the family of five at the front of the queue would all need new rail passes?

Translating the model from the assessment in your head, to the back of an envelope, to a spreadsheet, to a computer program, allows us to use the model to do science. Science starts with a hypothesis: a hypothesis is a simple model (see Section 1.4.1). An unambiguously stated hypothesis can be tested, allowing a comparison between two alternative theories. A numerically precise hypothesis can be quantified, allowing observations to be explained and future events to be predicted. Collecting hypotheses together allows their combined effects to be understood, and real environmental and ecological systems to be modelled. Rather than asking why we should use models, we should be asking why not!

Having started the modelling process by testing a qualitative hypothesis, it is a small step to make the hypothesis unambiguous and quantitative (see Section 1.4.1), adding value to our understanding. By measuring the time needed to serve customers at the railway station over a period of an hour, unambiguous quantitative hypotheses can be written to describe the movement of the queues. Consider the following example.

'On average, the time needed to serve a customer is:
   a seconds for a male customer aged less than 21;
   b seconds for a male customer aged from 21 to 65;
   c seconds for a male customer aged over 65;
   d seconds for a female customer aged less than 21;
   e seconds for a female customer aged from 21 to 65;
   f seconds for a female customer aged over 65.'

These hypotheses can be tested for this station by measuring the time needed to serve customers for a further period of time. If the new measurements are close to the hypotheses, I might decide to use them to predict which queue I should choose. This will need input information about the number of customers in the different categories in each queue. By doing this, I have written unambiguous and quantitative hypotheses, and I have quantified the expected result; this is modelling (although by the time I have done all this I would have missed my train and would possibly have been arrested!).

All of these measurements were taken at a railway station in London. When I am at a railway station in Paris, I bump into another modeller who has developed a different model for choosing which queue to join. His model is based on the following hypotheses.

'On average, the time needed to serve a customer is:

x seconds before 9.30 am;

y seconds between 9.30 am and 4.30 pm;

z seconds after 4.30 pm.'

We might use modelling to compare the two different theories (i.e., the time to serve the customer depends on the type of customer, or the time to serve the customer depends on the time of day). To settle our argument, further measurements should be taken at different railway stations both in London and Paris. The measurements should include the time of day, the number of customers in the different categories in each queue, and the time taken to reach the front of the queue. Examination of the measurements will put an end to all the arguments, and determine which model, if any, is likely to correctly predict the best queue to join.

Having collected together all the data needed to test the hypotheses, a small additional effort could give us so much more. We can use statistical analysis to determine any relationships that exist between the type of customer, the time of day and the time needed to serve the customer. We can determine the range of possible times for each queue, and so quantify the likelihood that our predictions are correct. Having quantified these relationships, we can try to understand the processes contributing to the observed result. If we understand the processes, we can extrapolate to railway stations in other countries: do the results hold equally well in England and France, and are they likely to be very different in North America? If we have established a good rule of thumb, I can tell my friends how to choose the best queue at any railway station around the world!

This example illustrates many capabilities that are common to most modelling activities.

In general, modelling has the potential to:

1. quantify expected results;
   - For example, given recent weather trends, do we expect sufficient rain to fall over the summer to keep reservoirs topped up, or should water companies be thinking about hosepipe bans?

2. compare the effects of two alternative theories;
   - For example, is organic food really better for our health, or is there no health benefit over non-organic foods?

3. describe the effects of complex factors, such as random variations in inputs;
   - For example, how does uncertainty in carbon dioxide emissions affect predictions of climate change?

4. explain how the underlying processes contribute to the observed result;
   - For example, describe how changes in the number of ragworms living in an estuary can cause changes in the populations of other organisms through complex food web interactions.

5. extrapolate results to other situations;
   - For example, what can the spread of BSE within the UK cattle herds tell us about the potential spread of BSE in France?

6. predict future events;
   - For example, in 2050 what will the global human population be?

7. translate our science into a form that can be easily used by non-experts.
   - For example, weather forecasts allow us all to make use of complex meteorological science.

All of these accomplishments are just a few easy steps from the starting point of the hypothesis.

Why use models on Mars?

Missions to Mars are expensive, but there have already been a number of missions to Mars, some landing robot probes on the surface,...

...others were secret manned missions to Mars — more about these later!

Models allow us to use the existing data to quantify expected results,...

...to compare the effects of two alternative theories,...

...to describe the effects of complex factors, such as random variations in input values,...

...to explain how the underlying processes contribute to the observed result,...

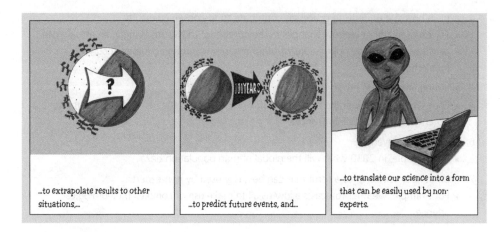

...to extrapolate results to other situations,...

...to predict future events, and...

...to translate our science into a form that can be easily used by non-experts.

1.4 Which model should I use?

Descriptions of mathematical models in the scientific literature often use a large number of different terms. It may appear that the only purpose of this jargon is to confuse. However, these terms can have one of two important purposes: to determine the type of model or to describe the type of mathematics to be used. Models and mathematics are often confused or equated to each other; but they are different things. The type of model tells us what the model does, that is, the sort of inputs used and outputs produced, the limitations of its application and the way it is used. The type of mathematics tells us how the model does that, that is, how it translates the inputs into the outputs. In the next two sections we attempt to demystify the jargon, using an example of a model of Martian fern to explain the purpose and meaning of some of the more commonly used terms. The importance of these terms will become clear throughout the book as we show how the type of model and mathematics needed can be used to guide the development process.

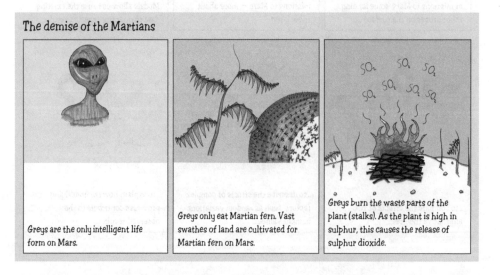

The demise of the Martians

Greys are the only intelligent life form on Mars.

Greys only eat Martian fern. Vast swathes of land are cultivated for Martian fern on Mars.

Greys burn the waste parts of the plant (stalks). As the plant is high in sulphur, this causes the release of sulphur dioxide.

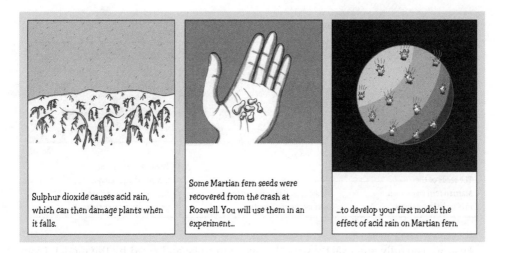

Sulphur dioxide causes acid rain, which can then damage plants when it falls.

Some Martian fern seeds were recovered from the crash at Roswell. You will use them in an experiment…

…to develop your first model: the effect of acid rain on Martian fern.

1.4.1 Determining what type of model to use

When we make efforts in modelling that go beyond the simple starting hypothesis, we need to define the type of results expected, and the type of model needed to achieve those results. If I want to tighten up a screw, I must first choose the screwdriver that fits the screw I am trying to tighten; I cannot use a flat-head screwdriver on a cross-head screw. In the same way, if I want a model to do a particular task, I must first choose the model that is appropriate to achieve that task. The classification of screwdrivers, by the shape and the size of the head, is extremely helpful when choosing the right screwdriver for the job. The classification of models, if it focuses on what the model can *do*, can also be extremely helpful when choosing the right type of model for the job.

The classification of models allows us to decide what type of model we need, and helps us to search for any models that have already been developed. There are many different ways of classifying models. In this discussion, we have started at the beginning, with the hypothesis. By examining the characteristics of the underlying hypotheses, all of the major features of models can be characterised, long before we start adding any numbers or equations.

Models can be based on anything from a simple hypothesis to a complex collection of hypotheses, but even the simplest hypothesis includes all the characteristics needed to classify a model. As an example, consider an experiment to study the effect of water acidity on the size of Martian fern.

Setting up a Martian fern pot experiment

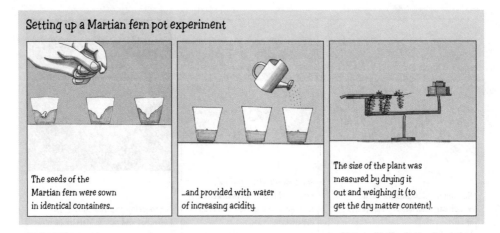

The seeds of the Martian fern were sown in identical containers...

...and provided with water of increasing acidity.

The size of the plant was measured by drying it out and weighing it (to get the dry matter content).

First, we carefully and exactly state the question to be addressed by the model. From the question addressed we can develop the **null hypothesis** and its alternatives. These will form the basis of our model.

The **null hypothesis** is a negative statement of the thing you want to test.
The **alternative hypotheses** list all possible alternatives to the null hypothesis.

The null hypothesis, referred to as H_0, for the research question 'Does the acidity of water have any effect on the size of Martian fern?' is

'*The acidity of water has no effect on the size of Martian fern.*'

The alternative hypotheses (H_1 and H_2) can then be formulated as follows:

H_1:   '*The size of Martian fern decreases with the acidity of water.*'

and

H_2:   '*The size of Martian fern increases with the acidity of water.*'

The model is then constructed from the hypothesis that is believed, from experimental observations, to be true.

Results of the Martian fern pot experiment

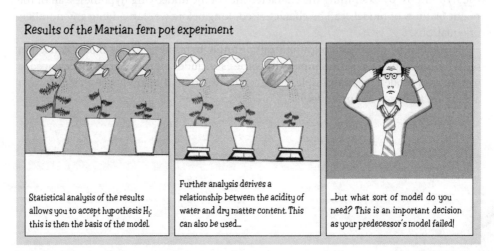

Statistical analysis of the results allows you to accept hypothesis H_1; this is then the basis of the model.

Further analysis derives a relationship between the acidity of water and dry matter content. This can also be used...

...but what sort of model do you need? This is an important decision as your predecessor's model failed!

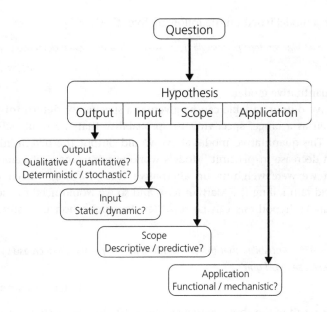

Figure 1.2 Classification of a model according to its underlying hypotheses.

The type of model that we should use can be defined by the **outputs**, **inputs**, **scope** and **application** of the hypotheses that it is based on (Fig. 1.2). The model will then be defined in the same way as the hypotheses.

The output from the above hypothesis describes the size of the Martian fern. The input is the pH and it describes the acidity of the water. The scope defines whether the model is to be used to explain the results of the current experiment, or whether it will be used to extrapolate to new situations. The application tells us what the model will be used for.

Outputs—information produced by the model.
Inputs—information needed by the model to run.
Scope—can the model be used outside the experiment used to develop it?
Application—is the model used to explain processes?

The **outputs** from the hypothesis could be **qualitative** or **quantitative**. A qualitative value describes the nature of the output, whereas a quantitative value will provide a numerical measurement or count. If we were interested in the more extreme effect, where acidic water kills the Martian fern, we might only need to know whether the fern was alive or dead (a qualitative measure):

H_1: *'The acidity of the water determines whether Martian fern is alive or dead'*

(Output: qualitative).

The input measurement of water acidity is quantitative, but the tested status of the fern is qualitative, that is, alive or dead. If a model provides any quantitative output, it is usually described as quantitative. A model based on the above hypothesis provides no quantitative output and so would be qualitative.

By contrast, a model based on the following hypothesis:

H$_1$: *'The size of Martian fern plants will decrease by 15% with each decrease in pH unit'*

(Output: quantitative)

would be a quantitative model.

A quantitative output variable can be given as a specific (or **deterministic**) value, or it can be given as a range, specifying the probability that the result falls within the given range. The quantitative model above would output the deterministic result of 15% for each decrease in pH unit. Models working with ranges of values are termed **stochastic**. If we were weighing up alternative risks, we might need to know the chances of acid rain killing the Martian fern, and so we would need to use a stochastic model. The above hypothesis can be rewritten for application in a stochastic model as follows:

H$_1$: *'There is 95% probability that the size of Martian fern plants will decrease by 10–20% with each decrease in pH unit'*

(Output: quantitative stochastic).

The stochastic result of the above model would be 10–20% for each decrease in pH unit.

The **inputs** to a model can be fixed values for each run of the model, or they can change over a series of measurements. The input measurement of the acidity of water does not change, and can be described as a **static** variable. However, we might be interested in how the size of Martian fern is affected by the acidity of water at different times from the start of the experiment. In this case, the inputs will change over a series of measurements taken at different times, and can be described as **dynamic** with respect to time. The hypothesis should be rewritten as follows to reflect our interest in time:

H$_1$: *'The size of Martian fern plants will decrease by 2% every day with each decrease in pH unit'*

(Input: dynamic with respect to time).

If a model uses any inputs that are dynamic, it is classified as a dynamic model. A model can be dynamic with respect to any of its inputs, such as successive changes in pH, but, in practice, the term 'dynamic' usually refers to changes with respect to time. Only if the model uses static input variables alone, will it be classified as static.

The **scope** of the hypothesis can be termed **descriptive** or **predictive**. A descriptive model is used to describe observations within the conditions of the current experiment. A predictive model is used to extrapolate beyond the scope of the experiment, and provides results that extend beyond the current observations. Models may be predictive with respect to time, space, species or any other input variable. If we wanted to use the above model to predict the effect of a higher acidity than included in the experiment, the model would be predictive with respect to acidity. A predictive model usually requires some degree of understanding of the processes causing change, as an understanding of the processes improves our ability to extrapolate to new situations.

The **application** can be **functional** or **mechanistic**. If the purpose of the hypothesis is to explain the underlying processes responsible for the overall result, the hypothesis can be described as mechanistic. If the hypothesis merely aims to represent or predict the experimental observations, the hypothesis will be described as functional. The inputs needed to drive a functional model are usually less complex than those needed to drive a mechanistic model. The need for mechanistic understanding in predictive models can result in confusion regarding the terms 'functional' and 'mechanistic'. A model that has been constructed around a mechanistic understanding of the underlying processes so that it can be used to predict, but that is used only to represent and predict the experimental observations, should be termed **functional**. This model should only be termed **mechanistic** if we are interested in using the model to find out what is happening and explain the underlying processes.

Failure due to choosing the wrong type of model

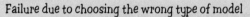

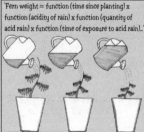

Your predecessor developed a model using data on Martian fern grown on sand.

He fitted a statistical model and used it to estimate the effects of acid rain across Mars.

The model was a *descriptive* model developed on sand. Clay soils were outside the development conditions.

He used it to try to predict the effects of acid rain on clay soils.

Clay releases aluminium under acid conditions. Aluminium is TOXIC to Martian fern.

The Martian fern died at higher pH than expected. The model FAILED. Acid rain was not controlled, Martian fern yields decreased and famine ensued.

SELF-CHECK QUESTIONS: WHAT TYPE OF MODEL SHOULD YOU USE?

1. Q: Which term best describes each of the following?
   a. Whether the model is predictive or descriptive?
      i. Inputs
      ii. Outputs
      iii. Scope
      iv. Application

   [A: iii. Scope]

   b. The values used by the model to predict the outcome?
      i. Inputs
      ii. Outputs
      iii. Scope
      iv. Application

   [A: i. Inputs]

   c. The type of relationships used in the model to explain observed results?
      i. Inputs
      ii. Outputs
      iii. Scope
      iv. Application

   [A: iv. Application]

2. You want to know whether a Martian fern is alive or dead at the end of an experiment in which it is exposed to acid water.

   Q: Is the output from the model qualitative or quantitative?

   [A: Qualitative—it is either alive or dead.]

3. You want to know by how much growth is reduced when the Martian fern is exposed to acid water.

   Q: Is the output from the model qualitative or quantitative?

   [A: Quantitative—the output should show the relative reduction in growth of plants grown using acid water compared to plants grown in neutral water.]

4. You want to know by how much each day the size of Martian fern plants decreases when exposed to acid water.

   Q: How would you describe the input variable, the acid water added each day?

   [A: Dynamic with respect to time.]

5. You use a statistical model to describe your results on the impact of acidity on Martian fern.

   Q: Is the scope of the model descriptive or predictive?

   [A: Descriptive—used within the same experiment from which the model was derived.]

6. You develop a model that simulates the effect of acidity on the growth of Martian fern using your results, which you then apply to fern growing in the wild.

Q: Is the scope of the model descriptive or predictive?

[A: Predictive—used outside the original data used to develop the model.]

7. You want to develop a model to predict what will happen to Martian fern under different possible future concentrations of acid rain. You develop the model from the relationships you found in an experiment where plant growth was measured under different acidities.

Q: Should the model application be described as functional or mechanistic?

[A: Functional—as it does not simulate how acidity affects plant growth.]

1.4.2 Determining what type of mathematics to use in the model

In Section 1.4.1, we described how the type of model is defined by the nature of the hypotheses that it is constructed from. The purpose of this section is to illustrate how the same question can be answered using many different types of mathematics. Many different mathematical approaches can be used to explore, explain and model the same data, and this adds more to the jargon of modelling and can be very confusing. However, the important distinctions between models do not lie in the type of mathematics used, but in the purpose for which the model was developed, that is, the type of model as described in Section 1.4.1.

The mathematics chosen are no more than the materials used to build the model. It is like choosing to build a house out of bricks, sticks or straw; different materials are more suitable for different purposes, but the important characteristic of the house is not the material it is made from, but rather the purpose it can be put to. The important characteristic of a straw house is that it is quickly-built temporary accommodation, not that it is made of straw. The important characteristic of a brick house is that it is strong and will not blow down, not that it is made of bricks.

However, because we know brick houses tend to be strong and do not blow away, a house is often described by the material it is made from rather than by its purpose. The same applies to models, so you will often hear a model being described by the mathematics it is constructed from. A model might be described as statistical, geostatistical, Bayesian, a neural network, cellular automata, process-based; the list goes on and on! As with the brick house, these terms become important only because we know something about the characteristics of models built using the different mathematical approaches. The characteristics of the approaches and how you choose an approach to develop your model will be discussed further in Chapter 2. Here, we discuss how the jargon describing the mathematical approach relates to the type of model and so helps to further describe its purpose. Do not worry about exactly how you should use each type of approach. Instead, focus on what the approach *does*, as it is this that allows you to assess what type of model can use this approach.

Many different mathematical approaches could be used to address the general research question 'How does acid rain on Mars affect Martian fern?' This question could be asked in many different ways; it is the exact form of the research question that determines the nature of the underlying hypotheses, the type of model associated with the hypotheses and the choice of mathematical approach.

If the question is 'How does acid rain on Mars affect the size of Martian fern?', then the *size* of Martian fern must be related to the pH of the rain water.

As already discussed in Section 1.4.1, this can be done through a hypothesis such as:

H_1: *'The size of Martian fern plants will decrease by 15% with each decrease in pH unit of the rain water at a chosen site.'*

The model developed from this hypothesis would be

- quantitative (because the output, *size*, is a quantitative measure),
- deterministic (because we are not specifically interested in risks so the simpler single-value result—*15% with each decrease in pH unit*—is sufficient),
- functional (because the model must describe *what* the effects of decreasing pH are, not how decreasing pH has this effect) and
- descriptive (because we do not intend to extend the results beyond the conditions of the original experiments).

As the model is descriptive and functional, there is no need for process-based equations; a **statistical** model is a good approach. A statistical model derives the mathematical equations used in the model directly from measured data using standard statistical techniques. The data could, for example, be analysed using an analysis of variance (ANOVA), as found in most statistical software packages. This is one type of mathematics that can be used to translate the hypothesis into a working model. When a model is described as statistical, you immediately know that it contains no understanding of the processes involved, so it is unlikely to be accurate if used to predict results well outside the conditions for which it was developed. Furthermore, you know that it can provide no mechanistic understanding of the observed responses because this information is not included in the statistical relationships derived from the data. The term 'statistical' immediately tells us that the model purpose *should* be only functional and descriptive.

If, however, you are interested in how the Martian fern is **distributed** with respect to the pH of the rain across Mars, the question would be reworded as 'How does the distribution of acid rain on Mars affect the distribution of Martian fern?' The hypothesis on which the model is based remains unchanged, but the research question dictates that the model now requires a spatial component to describe the spatial distribution. One set of mathematical approaches that can be used to describe the spatial distribution is termed **geostatistical**. To develop a geostatistical model, measurements of acid rain and Martian fern are collected at a series of points on a map. As with a statistical approach, the geostatistical procedures are used to fit the equations that determine the occurrence of Martian fern with respect to the pH of the rain water. These equations then form the body of the model. If a model is described as geostatistical, you know, as for the statistical approach, that it should only be used for functional and descriptive applications, but that it can also handle information about the spatial distribution of input data and provides a spatial distribution of results.

Alternatively, it might be the **risks** to Martian fern of the acid rain on Mars that are of most interest. In this case, the question becomes 'What are the risks to Martian fern of acid rain on Mars?' The hypothesis must be changed to express the risks associated with acid rain, for example:

H_1: *'The size of Martian fern plants will decrease by 10–20% with each decrease in pH unit of the rain water at a chosen site.'*

The model constructed from a hypothesis such as this will be quantitative, functional and descriptive (as before), but the outputs will now be stochastic, providing an idea of the risks associated with acid rain (a decrease of 10–20% with each decrease in pH unit). There are many different mathematical approaches for constructing such stochastic models.

Bayesian statistics incorporate prior knowledge and accumulated experience into probability calculations, based, for example, on the previous year's observations. Bayesian statistics could be used to address questions of risk. To do this, you would need to collect, in the previous year, measurements of the size of Martian fern and the pH of the rain water. From this information, the range of the likely decrease in Martian fern with the pH of the rain water can be ascertained. If a model is described as Bayesian, you know that it contains no understanding of the processes involved, and so should be used as a functional model, but that it can be used to predict risk.

A **neural network** has a similar function, and can be trained on the first year's data. A neural network is a piece of software that is 'trained' by presenting it with examples of input and the corresponding desired output. Neural networks mimic the vertebrate central nervous systems, to develop rules about relationships between inputs and outputs such as the decline of Martian fern with the increasing acidity of the rain. To use a neural network to assess the risks to Martian fern of decreases in the pH of rain water, measurements of Martian fern and rain water pH should be collected for a number of years and used to train the neural network. The neural network can then provide a very accurate representation of how the size of Martian fern has changed with changes in pH. A model constructed using a neural network is functional, but if sufficient data have been used to train then the model can be used to predict.

If, on the other hand, you know the source of acid rain, and want to know *how* the movement of acid rain from the point source affects the growth of Martian fern, then the question could be restated as 'How does the movement of acid rain from the point source affect the size of Martian fern?' The same hypotheses could be used to describe the effect of acid rain on Martian fern, but other hypotheses would be required to describe the movement of acid rain from the point source, such as the following:

H_1: *'The movement of acid rain is given by the direction and speed of the wind.'*

A number of different approaches exist for describing spatial changes in variables and could be used to construct a model of the movement of the acidic pollution from these hypotheses. The geostatistical approaches already discussed can be used in this context. Another approach is to use **cellular automata**. Cellular automata separate continuous space into discrete cells (see Fig. 1.3). The cells then react, by a series of rules or relationships, to the local conditions around the cell, for example, the condition of

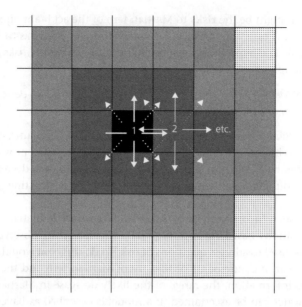

Figure 1.3 Using cellular automata to describe the movement of acid rain on Mars.

neighbouring cells. This creates a simple model describing the movement of the acidic pollution, allowing the amount deposited in rain in all cells around the point source to be calculated. The statistical relationship then provides an estimate of the effect of the acid rain in each cell on the Martian fern. If a model is constructed from cellular automata, you know that it describes movement across a region. As this requires a large number of calculations, the relationships used in the model tend to be simple statistical relationships, so cellular automata are usually functional models, providing process understanding of the movements only.

A very different approach is used if you want to know *why* the acid rain affects the Martian fern in the way it does. The research question must be restated as 'Why does acid rain on Mars affect Martian fern?' The simple hypothesis that is the basis of the statistical models must be replaced by a series of hypotheses describing the processes causing the Martian fern to be affected by acid rain water; hypotheses such as the following:

H_1: *'Aluminium is released from clay minerals according to the equilibrium constant for the acid reaction',*

H_1: *'Aluminium is toxic to Martian fern at concentrations of over 20 ppm',*

and so on.

The model constructed from this series of hypotheses is no longer functional; it is mechanistic because it explicitly contains information about the processes in the system. Models of this type are usually described as **process-based** models. If a model is described as process-based, you immediately know that it can be used to understand the mechanisms affecting the results, and, if the processes are adequately described, then it should be accurate when used to predict.

This discussion illustrates how the mathematics used to construct a model influence the capabilities and robustness of the model. Knowing the type of model needed will determine the type of mathematics that are appropriate; this helps in model development. Knowing the capabilities of the mathematics will determine the application and scope of an existing model; this helps us to decide if an existing model is appropriate for a given purpose. Far from being confusing, the jargon associated with models can greatly increase our understanding of models. The list of jargon presented here is not intended to be exhaustive; this is not feasible in a constantly developing field where different people use different names for the same thing. The examples given are intended to demonstrate how you should respond when you come across a new mathematical approach. A builder, faced with a new type of building material, will find out the characteristics of the new material (How heavy is it? How strong is it? What is its shape? How can it be fixed together?). The builder can then assess what type of house this new material can be used to build. Similarly, a mathematical modeller, faced with a new piece of jargon describing a mathematical approach, should find out the characteristics of the approach. If the modeller knows what type of model is needed, then the suitability of the new approach can be assessed more easily.

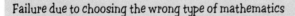

Failure due to choosing the wrong type of mathematics

Your predecessor developed a statistical model. The model fitted the results from the experiments growing Martian fern in sand...

...but when applied on a soil containing clay, the statistical relationships did not hold, because clay releases toxic aluminium as the soil acidifies. This was not in the model.

A mechanistic model that included a description of aluminium mobilisation at low pH would have been better able to predict Martian fern growth in a range of soils.

SELF-CHECK QUESTIONS: WHAT TYPE OF MATHEMATICS SHOULD YOU USE?

1. Q: To address the question 'Why does acid rain on Mars affect Martian fern?', which of the following mathematical approaches would most likely be inadequate?
   a. Mechanistic
   b. Statistical
   c. Neural network

   [A: b and c. Neither are process-based so cannot describe why acid rain affects Martian fern.]

2. Q: To address the hypothesis

    'The size of Martian fern plants will decrease by 15% with each decrease in pH unit of the rain water at a chosen site',

    which of the following mathematical approaches would most likely be inadequate?
    a. Mechanistic
    b. Statistical
    c. Neural network

    [A: None—all would potentially be capable of addressing the hypothesis.]

3. Q: For modelling spatial data, such as the distribution of Martian fern on Mars, which of the following mathematical approaches *could* potentially be used?
    a. Mechanistic
    b. Statistical
    c. Neural network
    d. Cellular automata

    [A: All could be used; d is ideally suited; a and b could be used if linked to a spatial data set of input data, and c could be used if enough spatial training data were available.]

4. Q: Which of the following are important when choosing the type of mathematics to use in a model?
    a. The question to be addressed
    b. The types of input data available
    c. Whether or not the mathematics is clever
    d. The hypothesis formulated from your research question

    [A: a and d are of primary importance; b also needs to be considered; and c is not important.]

1.5 Choosing an existing model

Before any time is wasted in developing an inappropriate model, the model required should be carefully classified using the above analysis. The model classification (whether static or dynamic, qualitative or quantitative, deterministic or stochastic, descriptive or predictive, or functional or mechanistic—see Section 1.4.1) will assist with searches of databases and the scientific literature for existing models in this area. You can then assess any available models for their suitability for your purposes. Developing your own model takes time and commitment, and like other areas of science should always build on the work of others. Carefully defining what is needed before you start will help you to do just that.

By classifying the type of model needed to answer the specific question, asked in a specific way, and understanding how the choice of mathematics relates to the type of model, you can quickly ascertain if an appropriate model already exists. If it does, and it is available to you, then you can try it out, and see if it does the job. Often, however, the right type of model does not exist (see Fig. 1.4). In this case, classifying the type of model

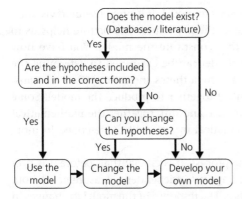

Figure 1.4 Should I develop my own model?

needed has saved you a huge amount of time learning how to use an inappropriate model, and wondering why it does not tell you what you want to know. If no suitable model exists then you have no choice but to make your own.

1.6 How is a model made?

Making a model that does just what you want it to do involves a series of important stages: model development, evaluation and application. These stages can be cumbersome and labour intensive, but if, in the interests of more rapid progress, you miss out any of these steps, then you are likely to end up with a model that does not behave itself and causes you embarrassment!

We start with the research question or practical objective (Fig. 1.5). It is valuable to write this down before you start to ensure that you do not get carried away with what

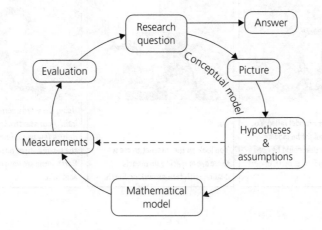

Figure 1.5 Stages in making a model.

you *could* model and forget what it is that you *need* to model. It is often useful to visualise the problem by drawing a picture to summarise the key issues. The picture helps us to list the main hypotheses and assumptions that constitute the model. You have now made a **conceptual model**. This conceptual model can be examined to determine the characteristics needed in the working model. The hypotheses and assumptions are transformed into mathematical equations and linked together to produce the model. Some of the many different approaches used to produce and link together the mathematical equations needed (i.e., to develop the mathematical model) will be described in more detail in Chapter 2.

Developing the model is not the end of the process. It is crucial that, before it is ever used in anger, the working model be evaluated against independent measurements from the range of situations in which it needs to work. The response of the model to changes in input variables should be determined using a sensitivity analysis and the results should be compared to the measured responses of these factors. The uncertainty associated with the simulations should be quantified, and an uncertainty analysis performed to quantify changes in uncertainty in different situations. This will be described further in Chapter 3.

Following successful model evaluation, you have confidence in your model, and can begin to use it. The responsibility for ensuring appropriate use of the model rests with the user interface. Complete tables of input defaults, comprehensive error and warning messages, full help support and system documentation, and the clear presentation of results, are all inherent in the art of model application, which will be discussed in Chapter 4. To enable us to apply models in the real world, we must understand how models are conceived, developed and tested. If you want to do this, read on!

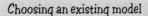

Choosing an existing model

Your predecessor's model failed. You wonder if there might be models already available that could be used to model Martian fern growth under acid rain...

You look on the Internet to see if there are any acid rain models dealing with fern growth on Earth.

Using the words 'fern', 'acid' and 'rain', your search engine turns up quite a few pages (over 247 000) on the sensitivity of fern to acid rain. Earth ferns are very sensitive to acid rain.

You refine your search with the term 'model' and the number of hits decreases—but there are one or two models out there.

You get copies of the models and try them out, but decide that none quite do what you want them to do and none have been tested on Mars! So you have to build your own model...

You will later build a model to do this, but you find that there are bigger problems associated with Martian fern cultivation that you will need to tackle first. Read on in Chapter 2!

■ SUMMARY

1. Models have the potential to
    a. compare the effects of two alternative theories,
    b. quantify expected results,
    c. describe the effects of complex factors, such as random variations in inputs,
    d. explain how the underlying processes contribute to the observed result,
    e. extrapolate results to other situations,
    f. predict future events, and
    g. translate our science into a form that can be easily used by non-experts.

2. Models can be classified by the hypotheses on which they are based.

3. A hypothesis can be classified by its outputs, inputs, scope and application.

4. A hypothesis can be classified by its outputs as
    a. qualitative (variable is a score or category), or
    b. quantitative (variable is a number).
    If a model includes any quantitative variables it is considered to be quantitative.

5. A quantitative hypothesis can be further classified as
    a. deterministic (variable is a single value), or
    b. stochastic (variable is the range of possible values).
    If a model includes any stochastic variables it is considered to be stochastic.

6. A hypothesis can be classified by its inputs as
    a. static (variables are fixed for any model run), or
    b. dynamic (variables change during the model run).
    If a model includes any dynamic variables it is considered to be dynamic.

7. A hypothesis can be classified by its scope as
    a. descriptive (interpolates observations), or
    b. predictive (extrapolates beyond observations).

8. A hypothesis can be classified by its application as
    a. functional (represents observations), or
    b. mechanistic (explains the underlying processes).

9. Models are further classified by the transformation of hypotheses into mathematical equations (e.g., statistical, geostatistical, Bayesian, neural networks, cellular automata, process-based).

10. Working models are made in the following three stages:
    a. development (research question leads to a conceptual model which leads to a mathematical model),
    b. evaluation (graphical, quantitative, sensitivity and uncertainty analysis), and
    c. application (user interface, defaults, error trapping, help support, presentation of results and system documentation).

■ PROBLEMS (SOLUTIONS ARE IN APPENDIX 1.1)

1.1. How would you classify the following model? A model has been developed to allow farmers to predict the potential increase in the weight of their cows with the amount of concentrate feed given each day. The model is structured around the following hypothesis:

> *'There is a 95% probability that the weight of cows will increase by 0.1–0.2 kg with each additional kilogram of feed given each day.'*

Specify the type of
a. outputs,
b. inputs,
c. scope and
d. application.

1.2. How would you classify the following model? A model has been developed to help policy makers estimate the changes in European soil carbon with changes in land management. This model must be used to quantify how soil carbon stocks might change in response to changes in land management and to determine improved management methods that might be used to increase soil carbon stocks. It is structured around the following hypotheses:

> Hypothesis 1: *'Land management (manure management, tillage practice and crop residue management) determines soil carbon stocks in cropland',*

> Hypothesis 2: *'Changing land management will change soil carbon stocks to a level determined by the new management regime.'*

Specify the type of
a. outputs,
b. inputs,

c. scope and
d. application.

1.3. **How would you classify the following model?** A model has been developed to help researchers estimate the size of a dolphin population, based on the probability of sighting a previously tagged dolphin on successive visits. The model is based on the following hypothesis:

'The probability of sighting a previously tagged dolphin is dependent on the size of the dolphin population.'

Specify the type of
a. outputs,
b. inputs,
c. scope and
d. application.

1.4. **How would you classify the following model?** A model has been developed to help bioenergy growers to determine which biofuel crops can be grown at different sites. The model is based on the following hypotheses:

Hypothesis 1: *'A biofuel crop will not grow outside a specified temperature range during the growing season'*,

Hypothesis 2: *'A biofuel crop will not grow above a specified elevation'*,

Hypothesis 3: *'A biofuel crop will not grow outside a specified rainfall range.'*

Specify the type of
a. outputs,
b. inputs,
c. scope and
d. application.

2 How to develop a model

Climate change on Mars — how can we reduce the loss of carbon from Martian soils?

Greys cultivate vast areas of land with Martian fern, their only food plant.

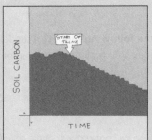

Cultivating the soil and harvesting the fern reduces the amount of carbon in the soil.

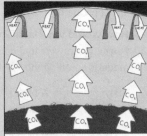

The carbon is lost to the atmosphere as carbon dioxide, a greenhouse gas that is causing global warming on Mars.

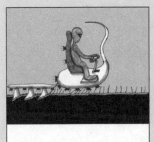

By reducing cultivation of the soil, more carbon could be kept in the soil, and less lost to the atmosphere.

More carbon could be built up in the soil by putting back organic waste from Grey sewage, composted with the waste parts of the Martian fern.

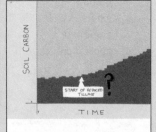

You need to know how much carbon could be gained in cultivated lands by reducing cultivation and adding more organic waste. You decide to develop a model!

Making a model involves three important stages: model development, evaluation and application. Model development is perhaps the least consistent of the three stages. Different modellers use very different approaches to devise their new models; some use very sophisticated procedures, while others develop models intuitively and give little thought to the stages they have used in the development. Here we attempt to provide one systematic methodology. Not all modellers use this approach, but this method should work until you develop your own favourite sets of procedures that are better suited to your own particular type of modelling. In our approach, we recommend that you choose an appropriate type of model for your purpose, draw up a conceptual model of this type, attach the mathematical relationships that make this into a numerical model, and then construct from this the computer model.

2.1 Choose the type of model

Choosing a model to help reduce carbon losses

Why am I doing this?   How should I do this?   What should I do now?

2.1.1 Why am I doing this?

Before you do anything, ask yourself why you are doing it. If you cannot answer that question, perhaps you should not be doing it at all; but hopefully a list of reasons for developing the new model will soon become apparent. The list of reasons will give you a number of important clues about the model you need to develop. It tells you what you want to know (outputs), what information is available to drive the model (inputs), and what you want to do with the model (scope and application). We have already discussed in Chapter 1 how the outputs, inputs, scope and application determine the type of model that you need.

2.1.2 How should I do this?

If we were to develop a model to estimate changes in soil carbon in the cultivated lands of Mars, the reasons for developing the model might be listed as follows:

1. to quantify how Martian soil carbon stocks might change in response to changes in agricultural land management; and
2. to determine improved management methods that might be used to increase Martian soil carbon stocks.

Reason 1 tells you that the output is the carbon stock in the soil of the cultivated lands on Mars. By the definitions given in Chapter 1, the output should be **quantitative** and **deterministic** unless very good reasons are later presented for making the model stochastic. The input is the current soil carbon stock of cultivated lands. Reason 1 also tells you that you need to use the model to predict the effects of land management change, so the scope is **predictive** with respect to land management, and land management is another input to the model. There is nothing to suggest that these inputs should be dynamic, so in the first instance the inputs will be **static**. Reason 2 determines the application of the model. You do not need to know *how* the carbon stocks change; at this stage you only need to know *by how much* it changes, so the application of the model is **functional**. You need to develop a quantitative, deterministic, static, predictive and functional model.

Having determined the type of model you need, now is the time to search the literature. If you find a model of the exact type needed, perhaps you can save a lot of time and effort by using what has already been done. If no model exists to do the job, then read on.

It is good practice, at this stage, to record the reasons for developing the model. It is easy to get carried away with what *can* be done with a model, and forget that the model does not actually do what you *need* it to do. Reasons for developing the model can change; as new issues are raised, old models are often reused for different purposes, but it is important that changes in the model function are recognised and properly examined rather than being quietly slipped in by the back door. Changes in application can demand further development or evaluation to ensure that the model still works well for the new application. The reduced availability of input data could require further functions to translate the limited data into inputs used by the model, but a reformulation might provide a better solution. If some values produced by the model are redundant, a simpler model might improve the reliability of simulations. These issues can only be properly addressed if the original reasons for developing the model have been recorded. The reasons for developing the model should provide a key paragraph in any final model documentation.

If, for example, having developed your model of changes in carbon stocks in soil, you give it to another scientist who is interested in soil carbon stock changes on a different, similar planet (such as Earth!), then the model would need to be adapted before it could be used. The possible changes needed are obvious when the reasons for developing the original model are considered. The model was developed to estimate changes in soil carbon stocks on Mars, using results from land management experiments on Mars; but do equations developed for Mars apply equally well on Earth? If not, equations should be re-derived using results from experiments on Earth. Are the soil carbon stocks of cultivated lands available for Earth? If not, the use of alternative input data should be considered. Without the list of reasons for developing the original model, another researcher might expect the model to work equally well under these very different conditions. When working with your own model, you might instinctively account for these changes, but if you give the model away these details can become lost. The list of reasons for developing the model together with the list of model assumptions and boundary conditions (discussed in the following sections) provide a systematic procedure that then allows models to be shared.

2.1.3 **The principle of parsimony**

In deciding what type of model to develop, you have strongly relied on one of the most important principles of modelling: the principle of **parsimony** or **Ockham's razor**.

> The principle of parsimony states: 'Of the two competing explanations, both of which are consistent with the observed facts, we regard it as right and obligatory to prefer the simpler' (Barker, 1961, p. 273).

The principle of parsimony tells us that the model developed should be no more complex than is absolutely necessary to describe the required outputs. In the example of the model of changes in soil carbon stocks, you rejected a stochastic, dynamic and mechanistic model because nothing suggested this extra complexity was needed to answer the posed questions. Dynamic models are generally more complex than static, quantitative more complex than qualitative, stochastic more complex than deterministic, predictive more complex than descriptive, and mechanistic more complex than functional.

Order of complexity in models

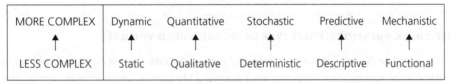

MORE COMPLEX	Dynamic	Quantitative	Stochastic	Predictive	Mechanistic
↑	↑	↑	↑	↑	↑
LESS COMPLEX	Static	Qualitative	Deterministic	Descriptive	Functional

When deciding if a model can be used for another purpose, the current model should only be retained if a simpler model could not provide the required outputs equally well; but why is this principle of parsimony so important?

The real world is very complex. In developing a model, the real world is summarised into simpler units. A good model should help us to understand changes occurring in the real world, even if that understanding takes some abstract mathematical form. If a process can be described equally well by two models, a simple and a more complex one, then the simpler model will tend to provide more understanding. The greater the number of units included in the model (where a unit could be a **parameter**, a **process** or a **sub-module**), the less explanation of the observations is afforded to each unit. Furthermore, the greater the number of units, the more likely it is that the individual units are dependent on each other, further impeding understanding.

All models include some component of error, and the units within the model all include their own component of error. If outputs from one unit act as inputs to another, any errors in the outputs from one unit are included as errors entered into the next; errors are said to **propagate** through the model. The greater the number of units, the greater will be the propagation of errors through the model.

A simpler model is a more useful model for the following reasons:

1. it provides greater understanding of the system as
   a. greater explanation is afforded to each individual unit, and

    b. individual units are less likely to be dependent on each other; and

2. it reduces propagation of errors through the model.

2.1.4 What do I do now?

Having decided *what sort* of model you need, the next step is to determine how the model will derive the output values that you *need* from the input values that you *have*. It is often useful to visualise the problem by drawing a picture to summarise the key issues and the connection between the inputs and the outputs. The picture helps us to define the bounds of the problem and list the main hypotheses and assumptions that constitute the model; this is termed a **conceptual model**. The conceptual model is particularly important as it gives us further hints as to when the model is no longer likely to work. It can then be examined to determine the characteristics needed in the mathematical model. The hypotheses and assumptions are transformed into mathematical equations and linked together to produce the final working computer model. The techniques needed to develop a conceptual model, a mathematical model and finally a computer model are described in the remainder of this chapter.

SELF-CHECK QUESTIONS: WHAT TYPE OF MODEL WOULD YOU USE?

1. You have data describing Martian fern growth over a year (total dry mass) and the mean annual temperature where it is growing. You want to know if Martian fern grows more in warmer areas.

    Q: Which of the following models would you choose?
    a. A dynamic mechanistic model of fern growth relating temperature to plant growth
    b. A static descriptive model of fern growth relating measured temperatures and total dry mass

    [A: b. The static descriptive model gives all of the information needed and is simpler.]

2. You want to develop a model to quantify how Martian soil carbon stocks might change in response to changes in agricultural land management.

    Q: Which of the following describes the type of outputs required for such a model?
    a. Qualitative
    b. Quantitative
    c. Deterministic
    d. Stochastic

    [A: b and c. The model should tell you how much soil carbon changes, and so should be quantitative. It should be deterministic because there are no good reasons to choose the more complex stochastic model.]

    Q: Which of the following describes the scope of the model needed?
    a. Descriptive
    b. Predictive

    [A: b. The model should predict how soil carbon changes with land management.]

3. You also want the model described in question 2 to determine improved management methods that might be used to increase Martian soil carbon stocks.

Q: Which of the following describes the application of such a model?

a. Functional

b. Mechanistic

[A: a. The model should tell you how much soil carbon changes, but not provide a mechanistic understanding of how land management changes soil carbon.]

4. Models vary in complexity.

Q: Why are simpler models often more useful than more complex models?

a. They often require less input data

b. Each component of the model explains more

c. Individual units are less likely to be dependent on each other

d. The propagation of errors is reduced compared to more complex models

[A: a, b, c and d. All are correct.]

2.2 Draw up a conceptual model

Drawing up a conceptual model to reduce carbon losses

You draw a picture to visualise the problem...so this is what you need to know!

From this you make a list of all the hypotheses and assumptions you need to include in the model,....

....and define which conditions the model should work for. This is your conceptual model!

In the literature, conceptual models can take any form from computer code with incomplete formulae, to a set of linked hypotheses, or even just a picture of boxes linked by arrows. In the context of mathematical modelling, we will introduce a more exact definition. A mathematical model is a simplified representation of reality described using mathematical formulae. A conceptual model is the simplified representation before the completed mathematical formulae have been attached. It is the framework that the mathematics will be hung on, so it is important to get it right. It defines the bounds of the problem, includes the complete list of hypotheses and assumptions needed to translate input values into outputs, and may also include some information about the form of the equations needed to describe each hypothesis. The conceptual model is the heart of the working model, but can be derived with no quantitative analysis of data; this is a happy coincidence that allows good models to be developed through the collaboration of scientists who know what they want but cannot translate it into numbers, and modellers who can translate information into numbers but are unsure as to the nature of the hypotheses. It also means that someone new to modelling can slip easily into good modelling practices without having to worry about the need for high-level mathematical skills. So if you are not a mathematics whiz-kid, take heart; many great modellers do not start their careers as mathematicians!

The list of reasons for developing the model is the place to start drawing up the conceptual model. In our example of changes in the soil carbon stock of cultivated lands on Mars, the model is being developed for the following reasons:

1. to quantify how soil carbon stocks might change in response to changes in land management; and

2. to determine improved management methods that might be used to increase soil carbon stocks.

2.2.1 Draw a picture

It often helps to visualise the problem by drawing a picture of what you are trying to do. This is especially true in a more process-based model, as the picture can depict the main processes that should be included, but it is even true in a very simple model, as it helps to focus the mind on the issues that are to be addressed. To draw a picture of the conceptual model, start with the information you have (the inputs), and end with the information you want (the outputs). In between, depict some method of getting from the inputs to the outputs. This may be a direct relationship, or may require a number of steps. In the model example, you know the soil carbon stock in cultivated lands on Mars at the current time, and want to determine how the soil carbon stock would change if different management practices were applied. You can draw a picture representing the inputs (present soil carbon stocks) and the outputs (future soil carbon stocks), and use a single arrow to indicate that you think you can derive a direct relationship between the inputs and the outputs (Fig. 2.1).

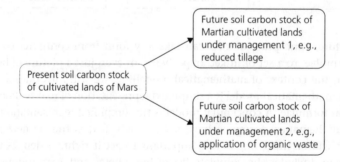

Figure 2.1 Conceptual model of changes in soil carbon in Martian cultivated lands under different management practices. The box with dotted lines on the left shows the input, the boxes with solid lines on the right show the outputs, and the arrows show the relationships between land management and change in soil carbon stocks.

2.2.2 List all hypotheses

The use of a single arrow to indicate that you think you can derive a direct relationship between present and future soil carbon stocks is built on the hypothesis that land management will affect soil carbon stocks in cultivated land, and that, by changing cultivation practice and organic waste management, present soil carbon stocks will change. The list of all the hypotheses to be incorporated in the model is the second important component of the conceptual model. This list determines which numerical equations must be derived when the mathematical components of the model are constructed; each hypothesis translates into at least one numerical equation.

Hypotheses

Hypothesis 1: *'Land management (cultivation practice and organic waste management) determines soil carbon stocks in cultivated lands on Mars.'*

Hypothesis 2: *'Changing land management will change soil carbon stocks to a level determined by the new management regime.'*

2.2.3 List all assumptions

Long before you try to attach the numerical relationships that quantify how each management practice will affect soil carbon, by representing the conceptual model in this way, you have made a number of assumptions. The conceptual model assumes that a single relationship between a land management practice and a change in soil carbon stock will apply for the whole cultivated area; all changes in soil carbon will occur similarly for all soil types, in all regions under all climate zones found on Mars. Although you have not specified how each management practice will affect soil carbon stocks, by drawing two output boxes, you have assumed that there will be different relationships between each management practice and the change in soil carbon. The list of these assumptions is the third important component of the conceptual model; it is especially important because it helps to set the **boundary conditions** of the model.

Assumptions

Assumption 1: 'The relationship between management and carbon stock is the same over all soils, regions and climate zones found on Mars.'

Assumption 2: 'The relationship between management and carbon stock is different for each practice.'

2.2.4 Set the boundary conditions

Setting boundary conditions builds into the model a complete set of modeller's excuses! The boundary conditions tell us the range of different conditions over which the model can be expected to simulate the result to within the required level of accuracy. Methods for calculating the accuracy of model performance are discussed further in the next chapter. The accuracy of model performance should, of course, be tested quantitatively for all the conditions under which the model is used. However, it saves time, improves user understanding and helps model testing if the developer defines, from the beginning, the conditions where logical understanding suggests that the model should work well. If subsequent testing of the model indicates a lower accuracy than expected within the stated boundary conditions of the model, then the assumptions have failed, and the model should be re-examined. On the other hand, low model accuracy outside the boundary conditions does not negate the assumptions or the model. If model testing indicates the failure of a model under specific conditions, and time, resources or knowledge prevent the model failure from being corrected, then the boundary conditions can be added to, so preventing model use in conditions where it is known to fail.

From the above list of assumptions, boundary conditions are immediately apparent. Assumption 1 states that the relationships are the same across Mars, so the spatial boundary of the model is the planet Mars. Assumption 2 states that the relationship is different for each practice, so the management boundaries are those practices that you derive equations for, namely, cultivation practice and organic waste management. The list of boundary conditions is the fourth important component of the conceptual model; it sets the scene for model testing and application, and it improves user understanding.

Boundary conditions

Spatial: Mars
Management: Cultivation practice
                     Organic waste management

You have constructed a simple conceptual model of changes in carbon stocks in the cultivated lands of Mars. You have summarised the problem as a picture, stated the full list of hypotheses and assumptions that are encapsulated within the conceptual model, and set initial boundary conditions for model application. You can now convert this conceptual model into a mathematical model by describing the relationships between land management and changes in soil carbon (shown by the arrows in Fig. 2.1) in terms of numerical equations.

SELF-CHECK QUESTIONS: WHICH CONCEPTUAL MODEL WOULD YOU USE?

1. You wish to build a model that is able to quantify how soil carbon stocks might change in response to changes in land management and determine improved management methods that might be used to increase soil carbon stocks.

   Q: Which of the following hypotheses should the model address?
   a. Land management (cultivation practice and organic waste management) determines soil carbon stocks in cultivated lands on Mars

b. Changing land management will change soil carbon stocks to a level determined by the new management regime

c. Land management changes the plant cover, which changes the temperature of the soil and affects decomposition

[A: a and b. Statement c may or may not be true, but in any case is unnecessary to address the questions you are interested in.]

2. The model described in question 1 lumps together data collected from all soils, regions and climate zones found on Mars to derive single relationships.

Q: What assumptions do you need to make for your simple model?

a. The relationship between management and carbon stock is different for different soils found on Mars

b. The relationship between management and carbon stock is different in different climate zones found on Mars

c. The relationship between management and carbon stock is the same over all soils, regions and climate zones found on Mars

[A: c. The model assumes that there is a single relationship between management and carbon stock across all soils, regions and climate zones found on Mars.]

3. The model is to be applied all over Mars for two management practices, namely, cultivation practice and organic waste management.

Q: What are the spatial boundary conditions of the model?

a. Each field of Martian fern

b. Each climatic region of Mars

c. Each soil unit on Mars

d. The whole of the planet Mars

[A: d. The whole of Mars.]

Q: What are the management boundary conditions?

a. Agricultural versus non-agricultural management

b. Cultivation practice

c. Organic waste management

[A : b and c. The two specific management practices considered define the boundary conditions.]

2.3 **Attach a mathematical model**

This is where things start to get technical, and where many potential modellers are frightened off! At this stage it can start to look more like mathematics or statistics than environmental or ecological science; but it is not. We are not in the business of inventing new mathematics or statistics. We are actually in the business of using old techniques that mathematicians and statisticians have already invented for us. What is more, software engineers have built many of the techniques we need into standard Windows programs. Modellers often find that they repeatedly need to use just a few of these methods; they become very adept at them, and do not worry about the rest. So do not be scared; mathematicians, statisticians and software engineers are here to help us! All we need to do is decide which technique to use, and that is what this section is about.

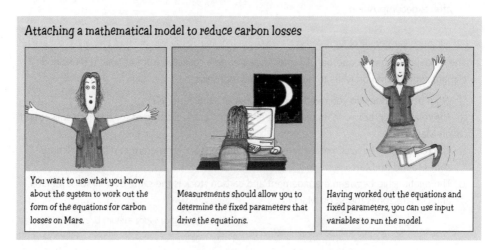

Attaching a mathematical model to reduce carbon losses

You want to use what you know about the system to work out the form of the equations for carbon losses on Mars.

Measurements should allow you to determine the fixed parameters that drive the equations.

Having worked out the equations and fixed parameters, you can use input variables to run the model.

2.3.1 The parts of a mathematical model

The mathematical model is made up of three parts. In modelling terminology these are often referred to as the following: **fixed parameters**, which, as the name implies, are values that, after being set to describe the thing you are modelling, then stay constant over different conditions; **input variables**, which are values that can change each time the model is run to describe the specific conditions of the run; and some way of **linking** together the fixed parameters and the input variables, which could be, for example, an equation or a neural network. Unfortunately, in statistics, input variables are usually described as **parameters**, which is confusing for modellers such as ourselves; in the modelling context we should remember that parameters are the things that *do not* change with each model run. Attaching the mathematics to our conceptual model requires the fixed parameters to be acquired in some way and linked to the input variables.

2.3.2 Linking together fixed parameters and input variables

There are three possible ways of getting at the fixed parameters and the link to inputs that is appropriate to the model. Firstly, we can derive a relationship between inputs and outputs directly from measured values, with no need to understand the processes causing the observed response. Secondly, we can examine what we know about the processes and propose an equation that should describe the measured data. Finally, or we can do both; we can use what we know about the processes to propose the form of the equation, but obtain the actual response by fitting the parameters in that equation to the measured data. The different approaches have different strengths and weaknesses. These make each approach suitable for a different purpose.

The range of techniques available to **derive a model directly from measured data** (e.g. by fitting an equation to measured data) should allow more variation in the data to be explained by this approach than by either of the other two. It is the best way to describe what we actually see because it is not constrained by our inevitably incomplete understanding of the processes. However, because it includes no description of the processes, we cannot be sure that the outputs will continue to respond in the same way beyond the bounds of our experiment. The application of the model is usually limited to the conditions of the experiment, and we should not use the model outside those conditions. In other words, this approach is usually used to develop **descriptive, functional** models.

Deriving a model from the scientific understanding of a system is an excellent way to test how good our understanding of a system actually is. If the model does not work, then we have got it wrong. If the model does work, then we know we have a very good understanding of the system, we can explain the reasons for observed changes, and we also know exactly where the model is likely to stop working. This approach may use a few measured values in the equations, but does not involve any fitting of parameters or equations to the data. However, in environmental and ecological science, we rarely have such a complete understanding of the system. Due to the complex and varied nature of ecology and the environment, fundamental rules are often more like guidelines than rules. The value of using these rules in a model is to provide an approximate solution, to set up a model structure that allows the rule to be tested for each specific case, and to suggest scientifically credible modifications to the rule that should improve the model accuracy. If model accuracy *is* improved, then we have learned more about the underlying processes.

If the solution to the equation is the thing you need in your model, then there is no more to be said about building the mathematical model; the equation is your model and you have finished. Often, however, the solution to the equation is just the starting point. You may be interested in how the solution will change with respect to another variable; in this case, you need to differentiate the equation with respect to the other variable. Alternatively, you may be interested in the change with respect to another variable that would produce the solution; in this case, you need to integrate the equation with respect to the other variable. Differentiation expresses the rate at which a variable changes with respect to the change in another variable on which it has a functional relationship. Integration is the process of finding the area under the curve that describes the relationship between the two variables. For more information, we refer you to some excellent basic texts on differentiation and integration, in the 'Further reading' section.

A model derived from scientific understanding is invaluable for highlighting where we have got it wrong, but might be too complex to use in many real-world applications as we often lack all of the input variables needed to drive the model. This approach will usually provide a **mechanistic** model that is **predictive** but **non-functional**.

Using the **two approaches together** incorporates strengths but also weaknesses from both. The model retains some degree of scientific understanding, so we can use it to explain what is going on, and we know where the model is likely not to work. However, the model derivation is constrained by our scientific understanding, so it will not usually describe the measured data as accurately as if the model was derived from the data alone. It may, however, be more accurate than a model derived purely from scientific understanding. This model is often better at predicting what is outside the conditions of the experiment (**predictive**) than a model derived purely from empirical data, it is good at explaining some of the underlying science (**mechanistic**), and it can sometimes also be applied with accuracy in the real world (so it is also **functional**).

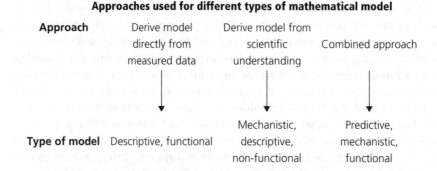

Approaches used for different types of mathematical model

Approach	Derive model directly from measured data	Derive model from scientific understanding	Combined approach
Type of model	Descriptive, functional	Mechanistic, descriptive, non-functional	Predictive, mechanistic, functional

2.3.3 Choosing the mathematical approach to derive the model

Modellers have used a wide range of different techniques to derive models, both directly from the measured data, and indirectly via model structures set up according to scientific knowledge. The choice of approach (i.e., the type of mathematics and the software used) is a subjective one, often dependent on the experience and knowledge of the modeller rather than on the nature of the data. This is a good thing, because it is by this mixing of different approaches to solve the same problem that new issues and scientific understanding emerge. We have listed some of these approaches below, and boxes describing the use of some of these approaches can be found on **Web link 2.1**, along with links to sites where you can find out more.

The form of the equation, the complexity of the system being modelled and the availability of data are the three factors that guide the choice of approach (see Fig. 2.2).

Form of equation: unknown. System: not complex. Data: limited

If you have no knowledge of the form of the equation and the system is not too complex, **statistical fitting** can be used even with only limited data (**Web link 2.1**, box 2.a). Standard spreadsheets such as Microsoft Excel (**Web link 2.2**) or Lotus 1-2-3 (**Web link 2.3**) can be used to fit the equations, and the model can be developed very quickly. With advances in software packages, this can be a very quick and easy process. For example, to fit an equation in Microsoft Excel, data is entered into the spreadsheet, and a curve can be fitted either by plotting the data and adding a *trendline*, or by using the *Analysis ToolPak* add-in. Both approaches take only a few minutes to complete.

Form of equation: unknown. System: complex. Data: extensive

If the system is more complex, standard statistical fitting procedures may not provide a very accurate model. If there is a large amount of data available to train the model, a **neural network** may provide a more accurate solution to the problem (**Web link 2.1**, box 2.b). There is a wide range of neural network software packages available, such as

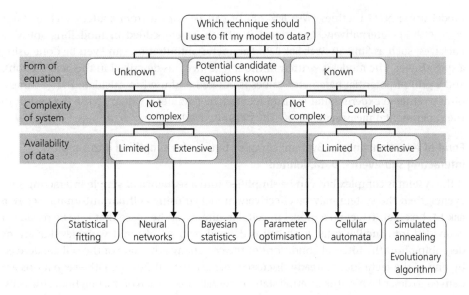

Figure 2.2 Which techniques could be used to develop your mathematical model?

SNNS (**Web link 2.4**), MathWorks (**Web link 2.5**) and NeuroSolutions (**Web link 2.6**). The data are entered into the package in sequence. Input nodes are set up, analogous to the neural nodes that exist in the brain. The input nodes are connected to output nodes via connections that are analogous to the nerve cells in the brain. This combination of nodes and connections is the neural network. All the rest of the work will be done for you by the software package. The results from the output nodes are compared to target outputs provided in the entered data. The goodness-of-fit between the calculated results and the target outputs is used to adjust the weightings given to each connection. Re-running the neural network gives a different result. Further correction of the connection weightings continues until a satisfactory simulation of the target outputs is achieved.

Form of equation: partially known

If a number of potential candidates exist for the form of the equation, then **Bayesian statistics** may provide a good modelling approach (see **Web link 2.1**, box 2.c). The software package WinBugs (**Web link 2.7**) is a useful tool for deriving such a model.

Form of equation: known. System: not complex. Data: limited

If the form of the equation is known from scientific understanding, and the system is not over-complex, then the fixed parameters that are contained in the equations can be derived by **parameter optimisation** (see **Web link 2.1**, box 2.d). These approaches simply adjust the values of the selected parameters so that the difference between the simulated and measured outputs is minimised. To do this you need a way of calculating the goodness-of-fit (see Chapter 3), a method for calculating the adjustment to the parameter values, and a procedure to iterate (repeat) the simulations until a satisfactory goodness-of-fit is achieved. The algorithms needed to do this can be embedded in the model using NAG routines (**Web link 2.8**) or can be copied from sources such as Press *et al.* (1991). Alternatively, the model itself can be embedded in modelling software packages, such as SimLab (**Web link 2.9**). Parameter optimisation can even be done using a spreadsheet. The model is written in the spreadsheet (as described in the next section), and a grid of potential parameter values is entered and used to calculate the goodness-of-fit to entered experimental measurements. The parameter values giving the best fit to the experimental measurements are the optimised values.

Form of equation: known. System: complex but can be simplified into simple interacting sub-events. Data: limited

If the system is complex but can be simplified into a sequence of simple interacting sub-events, then the system may be effectively modelled using **cellular automata** (see **Web link 2.1**, box 2.e). To set up a model in cellular automata, the space that is to be modelled is divided into smaller cells. Each cell is defined by a set of input variables. Equations describing how the internal conditions of the cell change the state of the cell are derived by one of the techniques already discussed (e.g., statistical fitting). Other equations are derived to describe how the internal state of the cell affects each of the neighbouring cells. These equations are then used to recalculate the state of every cell in the simulated space.

Form of equation: known. System: complex. Data: extensive

If the system is complex, and cannot be simplified by breaking it down into simple sub-events, but there is an extensive dataset available for developing the model, then the fixed parameters can be obtained by **simulated annealing** (see **Web link 2.1**, box 2.f).

Further optimisation of the actual form of the equations can be achieved using **evolutionary algorithms** (see **Web link 2.1**, box 2.g). These approaches must all be programmed using a language that can effectively handle large arrays of numbers, such as Fortran (**Web link 2.10**) or C++ (**Web link 2.11**).

2.3.4 **Example**

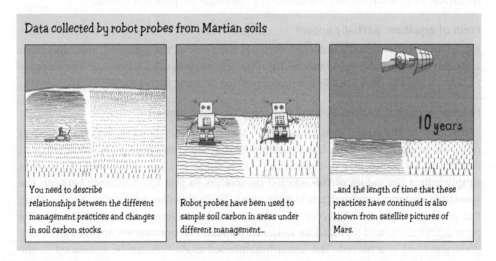

Data collected by robot probes from Martian soils

You need to describe relationships between the different management practices and changes in soil carbon stocks.

Robot probes have been used to sample soil carbon in areas under different management...

...and the length of time that these practices have continued is also known from satellite pictures of Mars.

10 years

There are many alternative approaches for adding equations to a conceptual model, but here we provide one example to show how it can be done. In this example we will take the model developed in Section 2.2 and add equations to illustrate how mathematics can be added to a conceptual model.

The conceptual model that you developed in Section 2.2 showed relationships (the arrows between the input and output boxes) between agricultural management and changes in soil carbon stocks on Mars. You know the type of model that you need from the reasons for developing it that you listed before you even formulated your conceptual model. The type of model tells you which approach is most appropriate. As discussed in Section 2.1, the model should be quantitative, deterministic, static, predictive and functional. The requirement for a predictive model suggests that you should derive the model from at least some degree of scientific understanding. The need for a functional model suggests that the equations should be fitted to the data. The combined approach should therefore be used to develop the model.

Scientific understanding suggests that a different relationship is required for changes in soil carbon stocks with each management practice. In the first instance, because the result is needed urgently, to prevent further loss of carbon dioxide to the atmosphere from soils on Mars, this might be the only use that is made of scientific understanding; everything else can be done by fitting. As you need to develop a model quickly, you might decide to quickly develop an imperfect model that works, rather than a superior model that takes years to complete. However, the possible limitations to prediction introduced by the chosen approach should be itemised alongside any model results, and the possibility of further, more scientifically based modelling should be considered.

The probe has sampled paired plots to compare soil carbon stocks under conventional tillage (ploughing) and reduced tillage (reduced soil disturbance). As the duration of the practice has varied, the measured difference in soil carbon can be divided by the duration of the treatment to get a yearly change in soil carbon under reduced tillage compared to conventional tillage. As the data are for only reduced versus conventional tillage, the best option is simply to take the arithmetic mean of these yearly changes to derive a crude estimate of how reduced tillage affects soil carbon stocks on Mars. Table 2.1 shows the data used to derive the arithmetic mean of yearly change in soil carbon under reduced tillage. For the data shown, the mean yearly change in soil carbon under reduced compared to conventional tillage is 0.4 tonnes of carbon per hectare (an area of 100 metres by 100 metres) per year (written as $0.4 \, \text{t} \, \text{C} \, \text{ha}^{-1} \, \text{y}^{-1}$).

The future soil carbon stocks under reduced tillage at a given time are given by the present soil carbon stocks plus the additional soil carbon that accumulates under reduced tillage over that time. The total change in carbon stocks due to changing management to reduced tillage will be roughly equal to this mean figure

Table 2.1 Results from paired samples of cultivated land on Mars taken by robot probes, showing changes in soil carbon stocks ($tC\,ha^{-1}\,y^{-1}$ to 30 cm soil depth) under conventional and reduced tillage.

Paired plot number	Duration (y)	Soil C stocks under conventional tillage ($tC\,ha^{-1}$ to 30 cm)	Soil C stocks under reduced tillage ($tC\,ha^{-1}$ to 30 cm)	Soil C increase through reduced tillage ($tC\,ha^{-1}$)	Yearly soil C increase through reduced tillage ($tC\,ha^{-1}\,y^{-1}$)
1	5	77.04	76.32	−0.72	−0.14
2	6	78.84	83.52	4.68	0.78
3	8	56.52	58.68	2.16	0.27
4	10	144	141.84	−2.16	−0.22
5	23	113.4	124.2	10.8	0.47
6	8	54.72	59.76	5.04	0.63
7	2	151.92	154.44	2.52	1.26
8	4	15.48	15.12	−0.36	−0.09
9	4	41.4	43.92	2.52	0.63
10	4	22.68	23.76	1.08	0.27
11	4	51.12	52.56	1.44	0.36
12	5	30.96	34.2	3.24	0.65
13	5	40.32	43.2	2.88	0.58
14	6	37.44	38.16	0.72	0.12
				Mean	0.40

multiplied by the number of years of reduced tillage. Thus we have the following relationship:

> Future soil carbon stock = Present soil carbon stock + (0.4 × Years under
> under reduced tillage zero tillage)
>
> $(tC\,ha^{-1})$ $(tC\,ha^{-1})$ $(tC\,ha^{-1}\,y^{-1} \times y)$

So the mean figure can be used to describe the relationship between present soil carbon stocks and future soil carbon stocks under reduced tillage, as indicated by the central arrow in the conceptual model shown in Fig. 2.1.

For organic waste management, a slightly more sophisticated approach can be used. As organic waste additions to the soil can vary in amount, relationships should be derived between the amount of organic waste added and the change in soil carbon. The robot probes were again used to sample paired sites on Mars where organic waste has been applied and those where it has not. This time the amount of organic waste added in each application and the number of years for which these applications were continued was obtained by hacking in to the computer records of the Greys. As the experiments have different durations, the measured difference in soil carbon should be divided by the duration of the treatment to get a yearly change in soil carbon. The data used to derive the yearly change in soil carbon for different organic waste addition rates are shown in Table 2.2.

The yearly change in soil carbon can be plotted against the manure addition rate. If you do this in a spreadsheet, you can use the facilities in the spreadsheet to fit a regression line to the graph and show the equation that describes the line. Figure 2.3 shows the plot of the data for organic waste addition.

The relationship between change in soil carbon ($t\,C\,ha^{-1}\,y^{-1}$) and the amount of waste added ($t\,DM\,ha^{-1}\,y^{-1}$; DM denotes dry matter) is given by the following regression equation (the equation of the line that is fitted through the data):

$$\textit{Annual change in soil carbon content} = 0.0145 \times \textit{Amount of organic waste added}$$
$$(t\,C\,ha^{-1}y^{-1}) \qquad\qquad (t\,DM\,ha^{-1}\,y^{-1})$$

As this equation tells us the change in carbon stocks due to added organic waste, it can be used in the same way as before, to obtain the relationship between present soil

Table 2.2 Results from paired plots on changes in soil carbon stocks ($t\,C\,ha^{-1}\,y^{-1}$ to 30 cm soil depth) with and without organic waste addition.

Paired plot number	Duration (years)	Yearly organic waste application rate ($t\,DM\,ha^{-1}\,y^{-1}$)	Soil C stocks with organic waste addition ($t\,C\,ha^{-1}$)	Soil C without organic waste ($t\,C\,ha^{-1}$)	Soil C increase due to organic waste addition ($t\,C\,ha^{-1}$)	Yearly soil C increase due to organic addition ($t\,C\,ha^{-1}y^{-1}$)
1	90	15	87.075	71.69	15.39	0.17
2	90	10	80.19	71.69	8.51	0.09
3	38	10.5	62.622	55.62	7.00	0.18
4	144	35	99.13	49.57	49.57	0.34
5	100	13.5	57.2	52.20	5.00	0.05
6	100	8	52.2	52.20	0.00	0.00
7	123	35	113.22	32.87	80.35	0.65
8	30	50	72.54	42.12	30.42	1.01
9	30	25	56.94	42.12	14.82	0.49
10	112	10	48.69	40.59	8.10	0.07
11	31	4.77	74.88	53.04	21.84	0.70
12	49	17.6	72	52.80	19.20	0.39
13	30	16.89	49.59	36.50	13.09	0.44
14	70	30	36.85	27.70	9.15	0.13
15	70	30	46.89	34.07	12.82	0.18
16	25	15	28.83	24.18	4.65	0.19
17	75	12	65.52	49.14	16.38	0.22
18	47	10	38.14	34.75	3.39	0.07

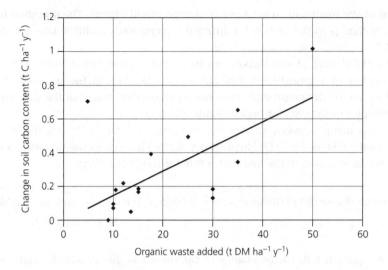

Figure 2.3 Relationship between the yearly amount of manure added and the yearly increase in soil carbon from data in Table 2.2. The fit is statistically robust ($R^2 = 0.366, p < 0.01$; see Chapter 3 for more details on evaluating your model).

carbon stocks and future soil carbon stocks under organic waste application, as indicated by the top arrow in the conceptual model shown in Fig. 2.1:

> *Future soil carbon stock under organic waste application* $=$ *Present soil carbon stock* $+$ *(0.0145× Amount of organic waste × Years of organic waste management)*
>
> (t C ha^{-1}) $\qquad\qquad (\text{t C ha}^{-1})$ $\qquad\qquad (\text{t C t DM}^{-1} \times \text{t DM ha}^{-1}\text{y}^{-1} \times \text{y})$

In this way, simple mathematical relationships can be attached to the conceptual model of the change in carbon content of cultivated soils on Mars. Very simple relationships have been used for this model, based on arithmetic means and simple regressions, to define the mathematical relationships needed. Remember that there is a range of possible techniques to derive the relationships that can be used in the model (see Fig. 2.2). Which technique you should choose depends on the nature of your model.

SELF-CHECK QUESTIONS: HOW WOULD YOU ATTACH A MATHEMATICAL MODEL?

1. You wish to derive a model that includes your best scientific understanding of the system you are studying.

   Q: Which of the following would be appropriate for developing your mathematical model?
   a. Derive the model directly form measured data
   b. Derive the model from scientific understanding
   c. Combine the approaches described in a and b

[A: b and c. Answer a is not appropriate as it simply fits to the data. Answer c is appropriate as it includes scientific understanding, even though it uses data to fit to equations.]

2. You wish to develop a simple model that tells you how tillage practice will affect soil carbon stocks on Mars. Your robot probe samples soils from land that has been ploughed and land that has not.

Q: What other information might you need to develop a simple model of the impact of tillage practice on soil carbon?

a. The names and locations of the Martian fields the samples were taken from

b. How long the tillage practices on the paired sites had been different

c. How much rainfall there had been on each field

[A: Only b is essential. a and c might be useful if developing more complex models, taking into account different regions or different factors, but the information is not essential at this stage.]

3. Your robot probe also samples fields that have either received, or have not received, regular additions of organic waste. You also know how much organic waste has been added to each field.

Q: What type of equation could be used to describe the relationship between the change in soil carbon and the amount of organic waste added?

a. Simple arithmetic mean of all the data

b. The median of all the data

c. The regression equation of the plot of the change in soil carbon versus the amount of waste added

d. An equation describing how soil carbon accumulates over time and how the amount of waste influences the change in soil carbon

[A: Approaches c or d are appropriate. You might not have enough data to describe how soil carbon accumulates with time for d, so you could use c (as in the example in this chapter). Approaches a and b would not take account of the different amounts of organic waste applied.]

2.4 Construct a computer model

Constructing a mathematical model to reduce carbon losses

You are pleased with your simple mathematical model, but to speed up calculations you need to build it into a computer model.

You build your model in Excel, but worry about the formulae being changed by the user.

So you learn to program in Fortran and program a stand-alone version of the model.

You have formulated a conceptual model. You have attached appropriate mathematical descriptions to make it into a mathematical model. Now you need to build this into a fully-functional computer model. How you do this depends on your needs. The options available are: general spreadsheets, which you may have already used in other aspects of your work; specialised modelling software, which software engineers have developed to make our lives easier; and high-level programming languages, which give us the greatest control over the models we develop.

To develop a computer model from the mathematical formulation you must do the following four things:

1. set aside computer memory to store the state of the system (often referred to as **state variables**);

2. tell the computer how one state affects another (as described by the mathematical equations);

3. obtain input variables from the computer user; and

4. pass output variables (the results) back to the computer user.

The software you use to develop the computer model helps you do these things. The selection of software is a balance between the effort required to develop the model and the ease of data entry, the presentation of results, the speed of simulation and computer memory requirements. If you have no experience of high-level programming languages, and are the only likely user of the model, then a general spreadsheet or specialised modelling software may be the quickest and easiest way to develop your model (although a proficient computer programmer will find a high-level language just as quick and easy). If, however, you are developing a computer model for wider use and application, it may be worthwhile investing the time to learn a programming language so that you can develop a model that is easy for other people to use. Further descriptions to help you decide between the available software are given below. Having decided what type of software to use, we list some sources of software and refer you to the specific software documentation for further information on its use.

2.4.1 General spreadsheets

A spreadsheet is a matrix of numbers and formulae. It usually displays data in grids (a series of columns and rows), ordered into worksheets (see, for example, the Microsoft Excel spreadsheet shown in Fig. 2.4). Each cell is referenced by the code along the top for the columns (letters) and the code down the side for the rows (numbers). You can use this code to refer to the cell. The cell can hold either a number or a formula that indicates how the value in the cell is derived from the values in other cells, referred to by their grid references. The data input into a cell is displayed so that the computer user can check that it has been correctly entered. The result, namely, the output from the calculation that is specified by the formula in the cell, is also displayed, so the user can check and record the result. In this way, a spreadsheet does for you all of the four things

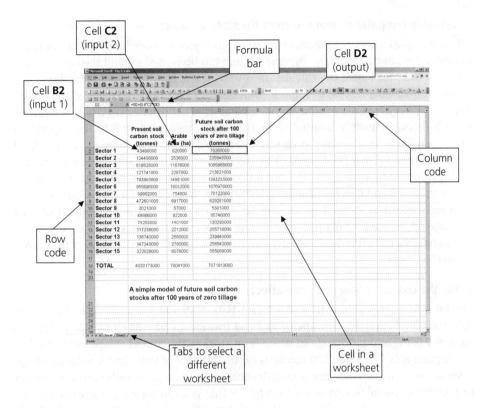

Figure 2.4 Constructing a computer model in Microsoft Excel.

that you need to do in order to translate a mathematical model into a computer model, namely, it sets aside computer space for the state variables (the data value in a cell), it tells the computer how one state affects another state (the formula in a cell), it allows data to be entered by the computer user, and it delivers outputs back to the user.

Part of the model of future carbon stocks in Martian cultivated land is shown in Fig. 2.4, constructed in Microsoft Excel. The following discussion describes how the facilities of a spreadsheet can be used to construct a working computer model. Microsoft Excel has been chosen to illustrate this because it is a very widely used computer package, so the details of Microsoft Excel application will be very familiar to most readers. These details are included here to help you to visualise explicitly what is being done when you set aside computer memory to store the state of the system, tell the computer how one state affects another, obtain input variables from the computer user, and pass output variables back to the computer user. This should then help you to better understand what is being described for the specialist software and high-level computer languages without the need for such specific descriptions. The key message of the following sections is not how to use Excel (for this you can refer to the Microsoft Excel user manual), but rather how the very familiar facilities of Excel relate to the modelling stages discussed.

1. Set aside computer memory to store the state variables

When the spreadsheet is opened, computer memory is set aside for the cells contained in worksheets. In this example, the first column has been used to hold the name of each region of Mars for which the calculation is to be done, and the first row has been used to hold the variable names. Column **B** contains the input variable, namely, the present soil carbon stock for each region. As the calculation is to be done on a regional basis, the result in hectares must be multiplied by the area of cultivated land in the region, and so the cultivated area (known from satellite remote sensing) is given in column **C**. All of this information has been typed in by the user, but could equally have been cut and pasted from another file using the normal Windows procedures. Column **D** is derived from columns **B** and **C** using a formula, in this case the formula for future carbon stocks with reduced tillage derived in Section 2.2. The formula for the highlighted cell, **D2**, is displayed on the formula bar. By entering values in the cells, computer memory of appropriate variable type and memory size has been set aside by Excel for columns **A** to **D** and rows **1–18**.

2. Tell the computer how one state affects another

The formula shown is **D2 = B2 + (0.4 * C2 * 100)**. Different spreadsheet packages have different conventions, but in Microsoft Excel this means **D2 = B2 + (0.4 x C2 x 100)**. The value 0.4 is the annual change in soil carbon stock with reduced tillage in t C ha^{-1}y^{-1} (see Section 2.2). The value 100 accounts for the number of years under reduced tillage.

Spreadsheets incorporate many facilities that make them ideal tools for simple modelling. One very useful feature of a spreadsheet is that you can copy and paste this formula into a new cell and the formula will change, depending on where the new cell is in the worksheet. So, if this formula was copied into the cell below it, cell **D3**, then the new formula would be **D3 = B3 + (0.4 * C3 * 100)**. By copying cell **D2** into all the other cells in column **D**, the calculation was completed for all of the regions on Mars. Similarly, if the cell was copied into **E2**, then the new formula would be **E2 = C2 + (0.4 * D2 * 100)**, although this would make no sense in the current example. This facility allows many different calculations to be performed very quickly, simply by entering a column or row of values and copy and pasting the formula into the adjacent column or row. If the entered values are sequential input variables (such as a sequence of time steps or measurements of weather), this facility can be used to create a dynamic model.

3. Obtain input variables from the computer user

Another familiar feature of a spreadsheet is the facility that allows parts of a formula to be held constant. This enables us to specify both the input variables and fixed parameters used by a model. If you do not want the formula to change, the spreadsheet includes symbols that tell it not to change the cell reference. In Microsoft Excel, this symbol is a dollar sign, '$'. If the dollar sign had been put in front of the row identifier giving the formula **D2 = B2 + (0.4 * C$2 * 100)**, then when it was copied into cell **D3** the formula would be **D3 = B3 + (0.4 * C$2 * 100)**, and in cell **E2** it would be **E2 = C2 + (0.4 * D$2 * 100)**. You would use this approach if the values given in row **2** were the input variables used to drive the model. If the dollar sign had instead been put in front of the column identifier, namely, **D2** = **B2** + (0.4 * **$C2** * 100), then when it was copied into cell **D3** the formula, namely, **D3** = **B3** + (0.4 * **$C3** * 100), would give the same result as without the dollar sign. In cell **E2** the formula would be **E2** = **C2** + (0.4 * **$C2** * 100). This approach would be used if the values in column **C** were the input variables. If the dollar sign was entered in front of both the row and the column identifier, namely, **D2** = **B2** + (0.4 * **C2** * 100), then no part of the cell identifier would change when it was moved. In cell **D3** the formula would be **D3** = **B3** + (0.4 * **C2** * 100), and in cell **E2** it would be **E2** = **C2** + (0.4 * **C2** * 100). This approach would be used if the value in cell **C2** was a fixed parameter that should remain the same in all parts of the model.

Input variables in row	⟶ Fix row identifier	(C$2)
Input variables in column	⟶ Fix column identifier	($C2)
Fixed parameter in cell	⟶ Fix row and column identifier	(C2)

4. Pass output variables (the results) back to the computer user

The output results are, of course, displayed in the cell that contains the formula. From here, results may be displayed on graphical plots, analysed statistically to determine goodness-of-fit to measured data (see Chapter 3), or exported to other software packages for further presentation and analysis. Together with the many other capabilities of spreadsheets, these facilities make it easy to create a model that responds to the entered data. Figure 2.5 shows example spreadsheets for simple models that are (a) qualitative, (b) deterministic, (c) stochastic and (d) dynamic. An exercise to construct the full computer model of carbon stocks in Microsoft Excel is given at the end of this chapter. The completed model is provided on the website accompanying this book (**Web link 2.12**).

Spreadsheets offer a number of important advantages to the modeller. It is usually very quick and easy to learn how to use a new spreadsheet package. Even if you have no experience of spreadsheets, given access to the documentation, you should be able to pick up enough expertise to create a new model in a few minutes. A spreadsheet allows you to see every stage of the calculation. This transparency can help to highlight where a model is going wrong. If you find a mistake, it is easy to quickly change the formula or input value and correct the model. Spreadsheets often include a wide range of extra tools that can help in model development, testing and the presentation of results; these include tools such as curve-fitting routines, tools for calculating statistics describing the accuracy of simulations, and graphical facilities that are compatible with word-processing packages and are of publishable quality.

The disadvantages of using spreadsheets relate to computer memory, speed of calculation, and the very easy access that is such an advantage for some applications. As described above, cells in the spreadsheet are allocated to describe the state, input and output variables for the model. The computer memory needed to hold the data in these cells is allocated permanently while the spreadsheet is open, rather than being temporarily allocated for the limited time that the variable is being used. This means that spreadsheet calculations tend to be very memory expensive. As the spreadsheet increases in size, the speed of recalculating cells, or even just moving around the spreadsheet, can become very slow. The spreadsheet can become too cumbersome to use if many formulae or data

(a)

(b)

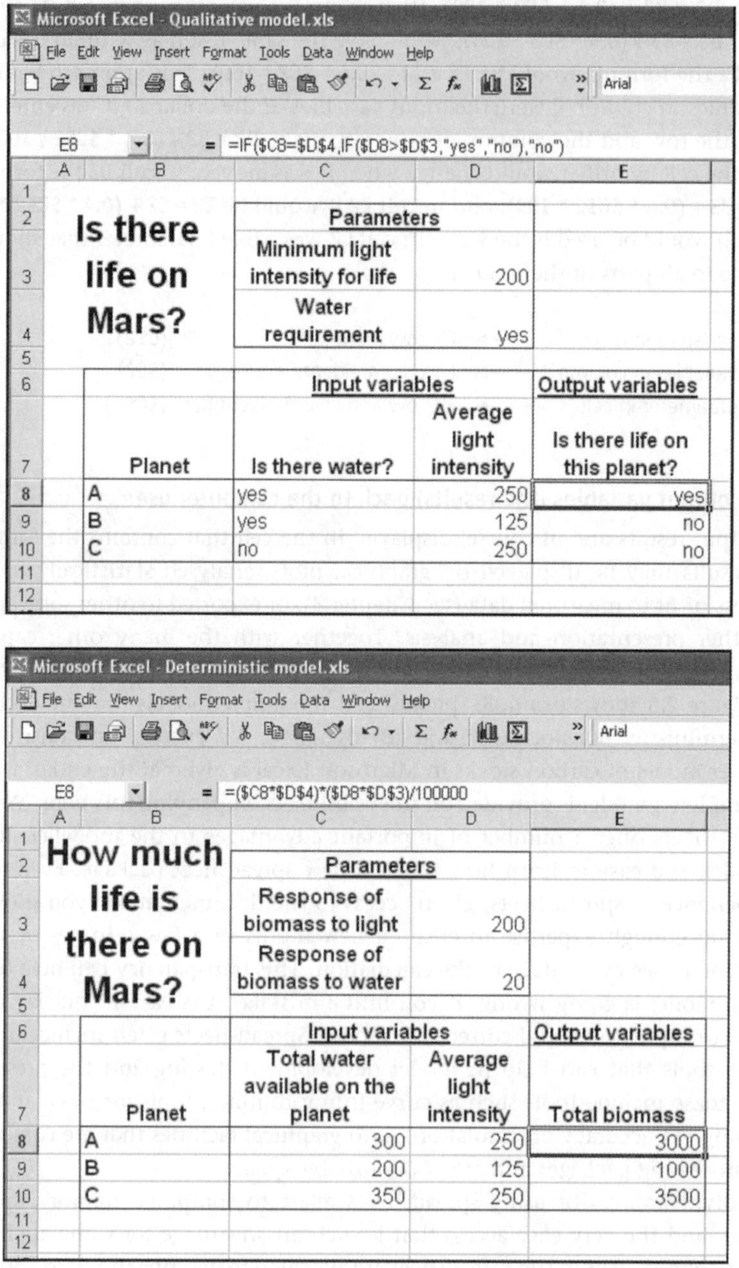

Figure 2.5 Different types of computer model in Microsoft Excel. (a) Qualitative model, 'Is there life on Mars?' showing the formula for the output cell **E8.** (b) Quantitative/deterministic model, 'How much life is there on Mars?' showing the formula for the output cell **E8.** (c) Quantitative/stochastic model, 'How much life is there on Mars?' showing the formula for the output cell **H8.** (d) Quantitative/dynamic model, 'How much life is there on Mars?' showing the formula for the output cell **F10.**

(c)

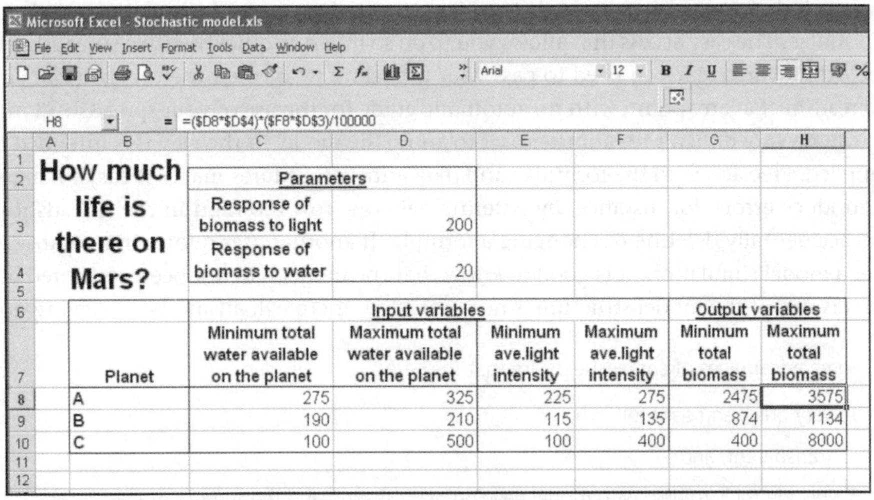

Microsoft Excel - Stochastic model.xls

H8 = =($D8*$D$4)*($F8*D3)/100000

How much life is there on Mars?

Parameters	
Response of biomass to light	200
Response of biomass to water	20

	Input variables				Output variables	
Planet	Minimum total water available on the planet	Maximum total water available on the planet	Minimum ave.light intensity	Maximum ave.light intensity	Minimum total biomass	Maximum total biomass
A	275	325	225	275	2475	3575
B	190	210	115	135	874	1134
C	100	500	100	400	400	8000

(d)

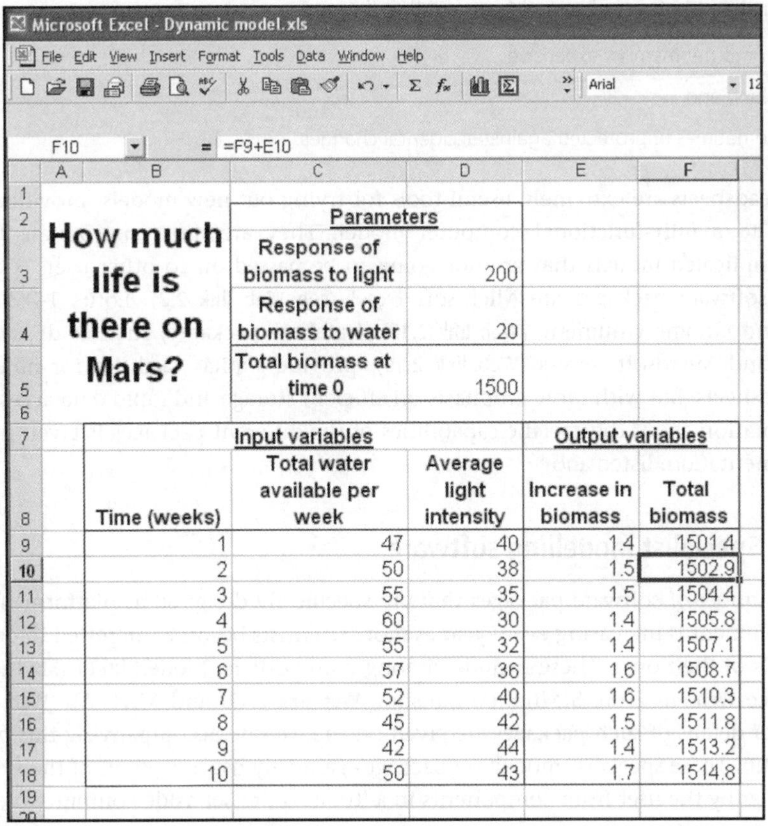

Microsoft Excel - Dynamic model.xls

F10 = =F9+E10

How much life is there on Mars?

Parameters	
Response of biomass to light	200
Response of biomass to water	20
Total biomass at time 0	1500

	Input variables		Output variables	
Time (weeks)	Total water available per week	Average light intensity	Increase in biomass	Total biomass
1	47	40	1.4	1501.4
2	50	38	1.5	1502.9
3	55	35	1.5	1504.4
4	60	30	1.4	1505.8
5	55	32	1.4	1507.1
6	57	36	1.6	1508.7
7	52	40	1.6	1510.3
8	45	42	1.5	1511.8
9	42	44	1.4	1513.2
10	50	43	1.7	1514.8

Figure 2.5 *(continued)*

are included. Spreadsheet models tend to be restricted to small simulations and simple formulae. The easy access that allows you to do so much so quickly with a spreadsheet can be a disadvantage if you need to pass your model on to another user. As you can enter a spreadsheet at any point, with no automatic guide for the use of your spreadsheet model, it can be very difficult for another user to apply the model in the way it is intended to be applied. Free access to the formulae and data entry procedures makes it easy for a user to introduce errors, for instance, by entering data one row too high in the spreadsheet, or by accidentally deleting or changing a formula. If another user quotes inaccurate results for a model simulation, it is good to know that the model had not been corrupted before it was run. If the model structure is not protected, there will always be a nagging doubt!

Some advantages of spreadsheets are that they are

1. very quick and easy to learn,

2. transparent, and

3. include tools to help with model development, testing and the presentation of results.

Some disadvantages of spreadsheets are that they are

1. computer memory expensive,

2. slow, and

3. sometimes unprotected against accidental changes.

Spreadsheets are extremely useful tools for trying out new models, providing a quick route to a fully-functional computer model. They are most appropriate for small, uncomplicated models that are not going to be passed on to other users. Commonly used software packages are Microsoft Excel (see **Web link 2.2**), Lotus 1-2-3 (**Web link 2.3**) and Gnome Gnumeric (**Web link 2.13**). Database packages, such as dBase (**Web link 2.14**) and Microsoft Access (**Web link 2.15**), provide similar facilities for modelling as spreadsheets, but with more emphasis on efficient storage and rapid data access. Further information about the specific capabilities of the different packages is given in the user documentation, listed above.

2.4.2 Specialist modelling software

The number of software packages that are specifically designed as platforms for model development is increasing every year as more scientists become interested in developing models of their own. These include, among many others, ModelMaker (SB Technology Ltd; see **Web link 2.16**), SIMILE (Simulistics; **Web link 2.17**) and MATLAB (**Web link 2.18**). Further details of such packages are given on the website accompanying this book (**Web link 2.19**). These specialist modelling packages generally use a diagram of the system, put together by the user from components in a 'toolbox', to set aside computer memory for the state variables, tell the computer how one state affects another, and exchange inputs and outputs with the user. This makes the steps between the conceptual model and the computer model much more intuitive, as the picture of the conceptual model and the diagram used to construct the computer model are usually very similar.

SB ModelMaker, a package that is typical of this type of software, is described by the developers as a model-building package for solving differential equations numerically (see Fig. 2.6).

1. Set aside computer memory to store the state variables

The user constructs the model from compartments that include numerical integration (square boxes), variables with no integration (round boxes) and defined values (angular boxes). This assigns computer memory to the state variables, as well as allowing the user to enter input variables and parameter values.

2. Tell the computer how one state affects another

The state variables are linked together by directional flows and influences, indicated by solid and dotted arrows, respectively. These arrows define the set of differential equations for the rates of change of state variables, telling the computer how one state variable affects another.

3. Obtain input variables from the computer user

Input variables are entered into the appropriate boxes by the user. This is a very transparent approach, which clearly displays how one variable might be expected to influence another.

4. Pass output variables (the results) back to the computer user

Having constructed an internally consistent diagram of the system, the model is run by pressing the 'Run' button. The output variables can then be passed back to the user either as a graph of one variable plotted against another, or as a table of results.

Figure 2.6 shows the example model of changes in carbon stocks on Mars with tillage, represented in SB ModelMaker. Software packages of this type include an increasing number of facilities that greatly improve the capabilities of modellers. SB ModelMaker provides an analysis of the model sensitivity to errors in a parameter. It provides functions to fit parameters to entered data. The entry of inappropriate input values can, to some extent, be guarded against, and will be highlighted by a warning message. With these growing libraries of modelling tools, specialist modelling software can be a good way of initially constructing a model, even if it is later re-coded in a high-level programming language.

Specialist modelling software has the advantage that it formalises the modelling process within one package, from the conceptual model (constructing the diagram from 'toolbox' components) to the mathematical model (fitting and optimisation procedures contained within the package) to the computer model. This may be too constraining for the purposes of some modellers, but it makes it easy to visualise the process and to understand each stage in the model development. The packages hold specialist tools that assist in every stage of modelling, from development to evaluation and application. Links to graphing and presentation facilities support the presentation and interpretation of results. The compilation of stand-alone software is possible, providing protection of algorithms and improved memory management over spreadsheets. Bug-free code is produced quickly and efficiently.

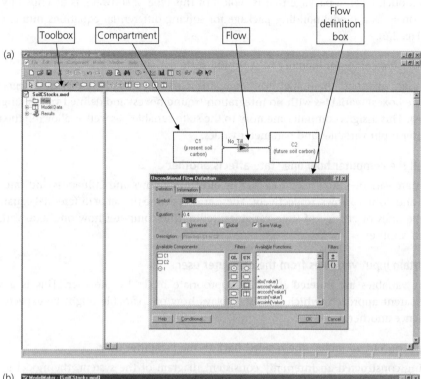

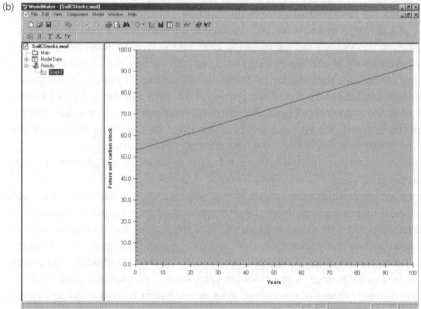

Figure 2.6 Constructing a computer model in SB ModelMaker. (a) Main model construction pane. (b) Graphical results pane.

The main disadvantage of such software can be that it constrains thinking into models constructed from compartments and flows. For some models, it is difficult to break the problem down into component parts used by the modelling software. For instance, ModelMaker can only solve equations of the form

$$\frac{dv}{dt} = \text{(algebraic expression)},$$

and so to solve more complex problems the developer needs a good grasp of differential equations. While it is possible to develop more complex models, as extra complexity is added, the modelling process can become cumbersome and time-consuming. In addition, because they are formulated around specialist concepts and functions, it can take more time to learn how to use these packages than the more general spreadsheet software.

Specialist modelling packages are becoming increasingly useful tools for modelling, providing a rigorous, formalised approach and a range of modelling tools. They can be used to develop more complex models than could be developed using spreadsheets, and can produce stand-alone code for application by other users. Further information about the specific capabilities of the different packages is given in the user documentation, listed above, and in the texts listed in the 'Further reading' section. Updated information is given on the website accompanying this book (**Web link 2.19**).

Some advantages of specialist modelling packages are that they

1. formalise the modelling process within one package,
2. provide specialist modelling tools,
3. provide access to graphing and presentation facilities, and
4. allow the compilation of stand-alone code.

Some disadvantages of specialist modelling packages are that they

1. constrain thinking into compartments and flows,
2. are more time-consuming to learn than spreadsheets, and
3. may be limited to less complex models.

2.4.3 High-level programming languages

High-level programming languages translate something approaching the English language into information that can be read by the computer (**machine code**). Machine code is incomprehensible to most of us, but can be efficiently read by the computer. The close-to-English, or **source code**, that you type in is **compiled** by the language software 'compiler' into **object** files, which are then linked together, usually into an **executable** file. The only part of this complicated process that you need to worry about is the executable file; the language software does the compiling and linking for you, with just a click of two buttons, namely, 'compile' and 'link'. The executable file (with file extension *. exe*) is a stand-alone program that can be run with a mouse click from

Windows, by typing the name of the . *exe* in a DOS screen, or by another user with no access to your original code. This gives your model protection from accidental changes by an inexperienced user, and prevents your ideas from being released before you are ready to publish them.

There are very strict rules governing the way that you enter your source code. Learning a programming language is simply learning what these rules are. This is no more complicated than learning the rules of writing a mathematical formula—'×' means multiply by, '+' means add together, and so on—although it can be more difficult as there tend to be more rules to learn. Some languages, such as Basic and Fortran, are very much easier to pick up than others, such as C++ and Java. The more difficult languages tend to give you better access to the computer memory; this makes them more efficient and more flexible, but also allows you to introduce some really nasty bugs!

1. Set aside computer memory to store the state variables

The source code that you write must explicitly set aside computer memory to store the state of the system as variables of different types (e.g., integers, real numbers or characters). Different types of variable use different amounts of memory. You can access these pieces of memory using the variable names that you have set up in your program.

2. Tell the computer how one state affects another

The way that one variable affects another can then be entered as an equation using the variable names.

3. Obtain input variables from the computer user

The input data can also be read into the correct place in the computer memory using these variable names. Data values can be entered by the user as values typed in after the program name (**arguments**), they can be read in from a data input screen, or they can be fed into the program from a file.

4. Pass output variables (the results) back to the computer user

The results of the model calculations can be output to a results screen, or they can be fed into an output file. Windows programming languages provide access to graphical user interfaces, allowing you to develop menu-driven input screens and automatic presentation of results in graphical plots.

There are a number of different computer languages to choose from. Basic is a simple language that is less structured than many other languages. This makes it easy to pick up the rules of the language, but means that the memory management is not as good as in more structured languages. A successor to Basic, Visual Basic (see Fig. 2.7(a)) includes access to Windows library functions and so is an excellent tool for creating Windows graphical user interfaces. This allows relatively inexperienced programmers to develop models with graphical user interfaces and screens to present the results, thus improving model sharing and allowing models to be constructed into applications such as decision support systems (see Chapter 4). Visual Basic is a good choice if you are an inexperienced programmer, if you need to share your model with other users, and if you are developing a relatively simple, memory inexpensive and quick-running model. Further information on Visual Basic programming is given in Wang (1998) and Foxall (2003), and on the World Wide Web at **Web link 2.20**.

Fortran (see Fig. 2.7(b)) has been the favourite language of scientists for many years. It is strongly structured into subroutines, so it is easy to write long and complicated models in Fortran. It is unforgiving if you break the rules governing how you enter the program code (**errors in format**). This is a good thing as it reduces the number of errors (**bugs**) introduced by mistyping. Originally, it did not allow direct access to the computer memory, which prevents some of the nastier crashing bugs that can haunt models. However, more recent Fortran compilers allow greater access to the computer memory. This is a useful addition, but should be used with care. Further information on Fortran programming is given in Etzel and Dickinson (1999) and Chapman (2003), and on the World Wide Web at **Web link 2.10**.

Languages such as C++ (see Fig. 2.7(c)) do allow direct access to the computer memory. This is very good if you know what you are doing, because it gives you full flexibility, allowing you to do whatever you can imagine with your computer. It also allows you to structure your program so that memory management is at its most efficient. It is, however, truly awful if you are an inexperienced programmer, as you can spend many frustrating hours trying to work out why your model keeps on crashing. If you are working with a complicated model that requires a large amount of memory so that you need to maximise the efficiency of your code, and if you are constructing it into an application for use by the wider public, then a programming language such as C++, or its Windows successor, Visual C++, is the only way to go; but it will come at a cost. It is more difficult to learn C++ than some of the other languages, and it is easier to introduce deep, hidden, memory-related errors. However, if you have the time, the inclination and the need to develop large, complex models supported by a graphical user interface, then Visual C++ is to be recommended. Further information on Visual C++ programming is given in Chapman (1998) and Davis (2000), and on the World Wide Web at **Web link 2.11**.

The reduced tillage component of the model of Martian soil carbon stocks, namely, $D = B + (0.4 \times C \times 100)$, is shown for comparison in Microsoft Visual Basic, Compaq Visual Fortran and Microsoft Visual C++ in Fig. 2.7. Notice that, at this very simple level, the programming code looks very similar in all three languages. Although strict rules govern the way you enter your source code, it is possible to understand the programs with no knowledge of the language at all! The rules of programming languages are usually intuitive, and so easy to learn. Figure 2.7(a) illustrates how visual programming languages help the programmer to develop code. Visual programming works through visual objects that are held in Windows libraries, for example, objects such as forms, buttons and text boxes. New objects are added to the project from a **toolbox** by clicking and dragging objects from the toolbox with the mouse. The properties of each object can be edited in the **properties window**. Through the properties window, the name of the object can be set, allowing you to access the computer memory where the object is stored, and enter the equations that make up the model. This visually-oriented approach will allow you to develop clear graphical user interfaces very quickly, although the manipulation of the underlying variables and formulae can be restrictive in Visual Basic.

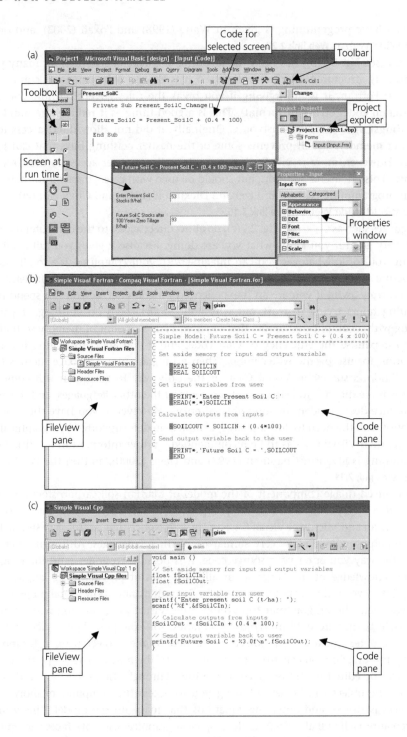

Figure 2.7 A program of a simple model. (a) Microsoft Visual Basic. (b) Compaq Visual Fortran. (c) Microsoft Visual C++.

Figure 2.7(b) illustrates the less visual, more code-oriented approach usually taken in Fortran. A list of the source files (the files holding the source code) included in the project is accessed in the **FileView pane**, and the selected source file is viewed in the **code pane**. At first sight, this seems more difficult to master than a more visual approach. However, many programmers find visual tools difficult to manipulate. If you have no need for a flashy graphical user interface, then a non-visual approach, using a clear, well-structured language such as Fortran, can offer the advantages of transparent and uncluttered programming.

A similar, code-oriented approach used in Visual C++ is shown in Fig. 2.7(c). The same visual toolbox is available in Visual C++ as in Visual Basic, but the more complex programs possible in C++ usually require a greater emphasis on freehand programming than does Visual Basic. Visual C++ offers the advantages of easy access to Windows library objects as well as non-restrictive manipulation of variables and formulae, and good memory management. The only disadvantage of Visual C++, as discussed above, is that it is more difficult to learn than the other languages. An exercise to construct the full model of carbon stocks in Visual Basic, Compaq Visual Fortran and Visual C++ is given in Problem 2.3 at the end of this chapter. The completed model is provided on the website accompanying this book (**Web links 2.21, 2.22 and 2.23**).

High-level programming languages offer a number of important advantages to the modeller. They give full flexibility over the type and complexity of model developed, and produce **stand-alone** software that does not require another software package to run. Input and output of data can be made highly user-friendly, guiding the user through the application of the model. However, it takes more time to learn a programming language than some of the other modelling tools already discussed. Programming and debugging the software can be time-consuming. This extra effort is worthwhile if you are developing a complex model that does not easily fit into the structures laid out in specialist modelling software, and is likely to be shared with another user.

Some advantages of high-level programming languages are

1. the flexibility of model structure and complexity,
2. the protection of source code,
3. efficient computer memory management,
4. stand-alone software packages, and
5. user-friendly input and output.

Some disadvantages of high-level programming languages are

1. the time needed to learn the programming language, and
2. the time needed to program and debug the source code.

This section should have provided sufficient information for you to decide if you need to use a high-level programming language to develop your model, and, if so, which one you should use. Further details on how to use the languages are given in the documentation listed in the 'Further reading' section.

SELF-CHECK QUESTIONS: HOW WOULD YOU CONSTRUCT A COMPUTER MODEL?

1. Q: Which of the following is not part of translating your model into a computer model?
   a. Set aside computer memory to store the state of the system
   b. Tell the computer how one state affects another (as described by the mathematical equations)
   c. Fix the parameters
   d. Obtain input variables from the computer user
   e. Pass output variables (the results) back to the computer user

   [A: c. Fixing the parameters is part of developing the model (Section 2.3), and not part of translating the model into a computer model.]

2. Q: Which of the following is the best method for converting your model into a computer model?
   a. General spreadsheets
   b. Specialised modelling software
   c. High-level programming languages

   [A: None is better than the other; each have their advantages and disadvantages.]

3. Q: What are the advantages when implementing your model on a computer using each of the following approaches?
   a. A spreadsheet
      i. It is easy to use
      ii. You can see what is going on
      iii. It runs quickly
      iv. It is suitable for very complex models
      v. It does not require highly-developed skills

      [A: i, ii and v. Spreadsheets are less suitable for very complex models and run more slowly than compiled computer programs written in high-level languages.]

   b. Specialist modelling software
      i. It is easy to use
      ii. You can see what is going on
      iii. It runs quickly
      iv. It is suitable for very complex models
      v. It does not require highly-developed skills

      [A: i and ii, and partly iii, iv and v. Specialist modelling software is intermediate between spreadsheets and high-level programming languages with respect to speed of running, the complexity of the models that can be easily created and the level of skill needed to use them.]

   c. High-level programming language
      i. It is easy to use
      ii. You can see what is going on
      iii. It runs quickly
      iv. It is suitable for very complex models
      v. It does not require highly-developed skills

      [A: iii and iv. High-level programming languages require highly developed programming skills, which means they are not easy to use relative to other options. Depending on the amount of output provided by the model, it may or may not be easy to see what is going on when the model runs.]

2.5 And then . . . .?

How you developed a model of changes in carbon stocks in Martian soil

You drew up a conceptual model...

...you got the data you needed from the robot probes, satellites and hacking, and derived the equations you would use...

...then you put the data and your equations into your computer and made a computer model. Are you ready to use the model?

We have worked through the example model of soil carbon stocks in cultivated lands on Mars, from the statement of the problem to the fully-functional computer-based solution. We have described how to decide what type of model to use from a list of the reasons for developing the model. We have worked through the representation of the problem as a conceptual model, namely, a picture of the system and the lists of hypotheses, assumptions and boundary conditions. We have shown how a mathematical model can be formulated around the conceptual model by deriving fixed parameters, input variables and the link between the input variables and the required results. Finally, we have described how the mathematical model can be developed into a computer model, using different types of software to set aside computer memory for the state variables, tell the computer how one state variable affects another, obtain inputs from the user, and return outputs to the user in an intelligible form. So now, can we get on and use the model?

We could use the model now, and it is very tempting to do so. However, before any model is applied in the real world, it should be rigorously tested. Using a model that has not been evaluated is like taking one measurement and using it to determine how the system behaves. The single measurement does give us some information, but without replicates we have no idea how accurate that measurement is. The measurement might have an error of 1% or an error of 200%. Clearly, the way that we use the measurement is very dependent on the error that we expect to be associated with the measurement. Similarly, model evaluation defines the error that can be expected in the model, and how we use the model is very dependent on the size of that error. The methods available to fully evaluate model performance are discussed further in the next chapter.

■ **SUMMARY**

A systematic method for model development might involve the following four stages.

- Stage 1: Choose an appropriate type of model.
- Stage 2: Draw up a conceptual model of this type.
- Stage 3: Attach mathematical relationships to the conceptual model.
- Stage 4: Construct from this a computer model.

Stage 1: Choose the type of model

1. List the reasons for model development.

2. Use the list of reasons to determine the simplest type of model needed.

3. Record the reasons for model development in the model documentation.

Stage 2: Draw up a conceptual model—a conceptual model is the list of hypotheses, assumptions and boundary conditions that define the model

1. Visualise the problem by drawing a picture.

2. Use the picture to list all the hypotheses needed to solve the problem.

3. Determine all the assumptions underlying the hypotheses.

4. List the boundary conditions for which the assumptions can be expected to hold.

Stage 3: Attach a mathematical model to the hypotheses of the conceptual model

1. The mathematical model consists of the following three parts:
   a. fixed parameters (which stay constant over different conditions);
   b. input variables (which change with each model run); and
   c. the link between input variables, fixed parameters and model outputs.

2. The link between input variables, fixed parameters and model outputs can be derived as follows:
   a. by fitting directly to measured values;
   b. from a scientific understanding of the system; or
   c. from a combination of scientific understanding and fitting.

3. Descriptive, functional models are usually derived by fitting directly to measured data.

4. Mechanistic, descriptive, but non-functional models are usually derived from scientific understanding alone.

5. Predictive, mechanistic and functional models are usually derived from a combination of scientific understanding and fitting.

Stage 4: Construct a computer formulation of the mathematical model

1. To develop a computer model you must do the following:
   a. set aside the computer memory to store the state variables;

b. tell the computer how one state affects another;

c. obtain input variables from the computer user; and

d. pass output variables back to the computer user.

2. This can be done using the following:

a. general spreadsheet packages;

b. specialised modelling software; or

c. high-level programming languages.

■ PROBLEMS (SOLUTIONS ARE IN APPENDIX 1.2)

2.1. **Construct a conceptual model** of the time taken to get served in a queue for tickets at the railway station.

a. List the reasons for developing the model.

b. Determine the type of model needed.

c. Draw a picture of the system.

d. List the hypotheses.

e. List the assumptions.

f. List the boundary conditions.

In the first queue, you observe that the average time to get served is about 60 seconds; in the second queue, this time is 90 seconds; while in the third queue, customers get served on average in 100 seconds. There are 10 people in the first queue, 7 in the second queue, but only 5 in the third queue. Construct a mathematical model of the three queues. Use a spreadsheet, a calculator or your head (!) to determine which queue you should join.

2.2. **Use statistical fitting procedures in a spreadsheet such as Microsoft Excel to derive a mathematical model** of the change in the height of trees with time after planting directly from the following data.

Time after planting (y)	Height of trees (m)	Time after planting (y)	Height of trees (m)	Time after planting (y)	Height of trees (m)
1	0.7	8	4.25	13.5	7.8
1.5	0.7	8.5	4	14	8
2	0.75	9	5.25	15	8.25
3	1	9.5	6	15.5	8.4
3.5	1	10	6	16	8.35
4	1.25	10.5	7	17	8.4
5	1.75	11	6.75	17.5	8.6
5.5	1.5	11.5	6.5	18	8.45
6	2.5	7.25	17.5	19	8.45
7	3.25	13	7.75	20	8.5
7.5	3.3				

2.3. **Carbon stocks on Mars change with the application of organic wastes** according to the following equation:

$$
\begin{array}{lll}
\text{Future soil carbon} \\
\text{stock under organic} \ = \ \text{Present soil carbon stock} \ + \ \begin{array}{c} (0.0145 \times \text{Amount of organic} \\ \text{waste} \times \text{Years of} \\ \text{organic waste management}) \end{array} \\
\text{waste application}
\end{array}
$$

$$
(\text{t C ha}^{-1}) \qquad\qquad (\text{t C ha}^{-1}) \qquad\qquad (\text{t C t DM}^{-1} \times \text{t DM ha}^{-1}\text{y}^{-1} \times \text{y})
$$

The following data's for use in the calculations.

	Present soil carbon stock (t C)	Arable area (ha)	Average application rate of organic manure for the sector (t DM ha^{-1} y^{-1})
Sector 1	43 460 000	820 000	50
Sector 2	134 408 000	2 536 000	15
Sector 3	618 828 000	11 676 000	25
Sector 4	121 741 000	2 297 000	40
Sector 5	793 993 000	14 981 000	10
Sector 6	955 696 000	18 032 000	0
Sector 7	39 962 000	754 000	20
Sector 8	472 601 000	8 917 000	2
Sector 9	3 021 000	57 000	7
Sector 10	48 866 000	922 000	8
Sector 11	74 253 000	1 401 000	7
Sector 12	117 236 000	2 212 000	2
Sector 13	136 740 000	2 580 000	3
Sector 14	147 340 000	2 780 000	50
Sector 15	322 028 000	6 076 000	25
Total	4 030 173 000	76 041 000	

Use the data in the table above to determine the total changes in carbon stocks over the next 10, 20, 30, ..., 100 years, assuming organic manure applications remain unchanged, and carbon stocks continue to increase with continued manure additions. Which sector will have the highest carbon stock after 50 years of manure applications? Which sector will have the highest carbon stock after 100 years of manure applications?

a. **In a general spreadsheet, such as Microsoft Excel**, construct the complete model of the carbon stocks of Martian land with organic waste application.

b. **If you can program in Visual Basic**, construct the complete model of the carbon stocks of Martian land with organic waste application in Visual Basic.

c. **If you can program in Fortran**, construct the complete model of the carbon stocks of Martian land with organic waste application in Fortran.

d. **If you can program in C++**, construct the complete model of the carbon stocks of Martian land with organic waste application in Visual C++.

■ FURTHER READING

Microsoft Excel

Frye, C. (2003). *Microsoft Office Excel 2003 step by step*. Microsoft.
  (*Hands-on, self-study guide, building the skills you need to succeed with Excel.*)

Stinson, C. and Dodge, M. (2003). *Microsoft Office Excel 2003 inside out*. Microsoft.
  (*Conquer Excel 2003—from the inside out.*)

Winston, W. L. (2004). *Microsoft Excel data analysis and business modelling*. Microsoft.
  (*Discover data analysis and modelling techniques.*)

ModelMaker

Bernard, M. (1995). ModelMaker 2.0 for Windows. *Biotechnology Software Journal*, **12**, (6), 16–19.

Blocher, M., Walde, P. and Dunn, I. J. (1999). Modeling of enzymatic reactions in vesicles: The case of alpha-chymotrypsin. *Biotechnology and Bioengineering*, **62**, 36–43.

Citra, M. J. (1997). ModelMaker 3.0 for Windows. *Journal of Chemical Information and Computer Sciences*, **37**, 1198–200.

David, G. (1999). Mitochondrial clearance of cytosolic Ca2+ in stimulated lizard motor nerve terminals proceeds without progressive elevation of mitochondrial matrix [Ca2+]. *Journal of Neuroscience*, **19**, 7495–506.

Lundin, D. (1987). Ruminations of a model maker. *IEEE Computer Graphics and Applications*, **7**, (5), 3–5.

Martin, S. (1998). A population model for the ectoparasitic mite Varroa jacobsoni in honey bee (Apis mellifera) colonies. *Ecological Modelling*, **109**, 267–81.

McKellar, R. C. (1997). A heterogeneous population model for the analysis of bacterial kinetics. *International Journal of Food Microbiology*, **36**, 179–86.

Meek, P. (1997). ModelMaker for Windows (Version 2.0). *Addiction Biology*, **2**, 240–1.

Müller, C. (1999). *Modelling soil–biosphere interactions*. CABI Publishing, Wallingford.

Negro, C. A., Hsiao, C. F., Chandler, S. H. and Garfinkel, A. (1998). Evidence for a novel bursting mechanism in rodent trigeminal neurons. *Biophysical Journal*, **75**, 174–82.

Sands, P. J. and Voit, E. O. (1996). Flux-based estimation of parameters in S-systems. *Ecological Modelling*, **93**, 75–88.

Selzer, P. (1998). ModelMaker—System modeling on a PC. *Nachrichten Aus Chemie Technik Und Laboratorium*, **46**, 652–6.

SB Technology (1995). *ModelMaker for Windows 's, (Version 2.0)*. Cherwell Scientific Publishing, Oxford.

Starr, M. (1999). WATBAL: A model for estimating monthly water balance components, including soil water fluxes. In *8th annual report 1999 UN ECE ICP integrated monitoring* (ed. S. Kleemola and M. Forsius). The Finnish Environment 325, pp. 31–5. Finnish Environment Institute, Helsinki.

Thomas, P. and Waring, D. W. (1997). Modulation of stimulus–secretion coupling in single rat gonadotrophs. *Journal of Physiology (London)*, **504**, 705–14.

Viola, R. (1996). Hexose metabolism in discs excised from developing potato (*Solanum tubersosum* L.) tubers. II. Estimation of fluxes *in vivo* and evidence that fructokinase catalyses a near rate-limiting reaction. *Planta*, **198**, 186–96.

Watson, A. (1998). ModelMaker for Windows (Version 3). *Chemistry in Britain*, **34**, 54.

Wimpenny, J. (1995). ModelMaker 2.0 for Windows. *Binary Computing in Microbiology*, **7**, 4–6, 113–14.

Visual Basic

Foxall, J. (2003). *SAMS teach yourself Microsoft Visual Basic. NET 2003 in 24 hours.* SAMS Publishing, Indianapolis, IN.
(*A straightforward, step-by-step approach to learn the essentials of Visual Basic. NET.*)

Wang, W. (1998). *Visual Basic 6 for dummies (for Windows).* Wiley, Chichester.
(*A witty, well-written guide to Visual Basic 6.0.*)

Fortran

Chapman, S. J. (2003). *Fortran 90/95 for scientists and engineers* (2nd edn). McGraw-Hill, Boston, MA.
(*Teaches Fortran in a style suitable for use on large projects, emphasising the importance of going through a detailed design process before any code is written.*)

Etzel, M. and Dickinson, K. (1999). *Digital Visual Fortran programmer's guide.* Digital Press, Burlington, MA.
(*Details the process of developing Digital Visual Fortran Version 6 applications, offering capabilities and coding guidelines for each project type.*)

Visual C++

Chapman, D. (1998). *SAMS teach yourself Visual C++ 6 in 21 days.* SAMS Publishing, Indianapolis, IN.
(*Covers all the essentials of basic Windows and Microsoft Foundation Classes (MFC) development, and addresses new features in Visual C++ 6.*)

Davis, S. R. (2000). *C++ for dummies* (4th edn, completely revised). Wiley, Chichester.
(*How to write programs, create source codes, use the Visual C++ help system, build objects, develop C++ pointers, debug programs, and more.*)

Integration and differentiation

Wainwright, J. and Mulligan, M. (2004). *Environmental modelling. Finding simplicity in complexity.* Wiley, Chichester.

Parameter optimisation

Press, W. H., Flannery, B. P., Teukolsky, S. A. and Vetterling, W. T. (1991). *Numerical recipes in C. The art of scientific computing.* Cambridge University Press.

Neural networks

Gurney, K. (1996). *An Introduction to neural networks.* UCL Press, London.

Smith, L. (2003). *An introduction to neural networks.*
http://www.cs.stir.ac.uk/~lss/NNIntro/InvSlides.html

Stergiou, C. and Siganos, D. (2004). *Neural networks.*
http://www.doc.ic.ac.uk/~nd/surprise_96/journal/vol4/cs11/report.html

▓ REFERENCES

Barker, S. (1961). The role of simplicity in explanation. In *Current issues in the philosophy of science: Proceedings of the American Association for Advancement of Science, 1959* (ed. H. Feigl and G. Maxwell), pp. 265–74. Holt, Rinehart, and Winston, New York.

▓ WEB LINKS

Web link 2.1: Online resource centre: **www.oxfordtextbooks.co.uk/orc/smith_smith/**

Web link 2.2: **http://office.microsoft.com/en-us/FX010858001033.aspx**
Microsoft Excel home page

Web link 2.3: **http://lotus.com/products/product2.nsf/wdocs/123fact**
IBM Lotus 1-2-3 factsheet

Web link 2.4: **http://www-ra.informatik.uni-tuebingen.de/SNNS/**
Neural network software package

Web link 2.5: **http://www.mathworks.com/products/neuralnet/**
MathWorks neural network toolbox

Web link 2.6: **http://www.nd.com/**
NeuroSolutions 4.3—neural network modelling software

Web link 2.7: **http://www.mrc-bsu.cam.ac.uk/bugs/welcome.shtml**
WinBugs

Web link 2.8: **http://www.nag.co.uk/**
NAG routines for parameter optimisation

Web link 2.9: **http://www.simlab.de/**
SimLab home page—parameter optimisation software

Web link 2.10: **http://www.qtsoftware.de/dvf/index2.html**
Compaq Visual Fortran home page.

Web link 2.11: **http://msdn.microsoft.com/visualc/**
Microsoft Visual C++ Developer Center

Web link 2.12: Online resource centre: **www.oxfordtextbooks.co.uk/orc/smith_smith/**

Web link 2.13: **http://www.gnome.org/projects/gnumeric/**
Gnome Gnumeric spreadsheet home page

Web link 2.14: **http://www.dbase.com/**
dBase home page

Web link 2.15: **http://office.microsoft.com/en-us/FX010857911033.aspx**
Microsoft Access home page

Web link 2.16: **http://www.modelkinetix.com/modelmaker/**
ModelKinetix ModelMaker introduction

Web link 2.17: **http://www.ierm.ed.ac.uk/simile/**
SIMILE home page

Web link 2.18: **http://www.mathworks.com/**
MathWorks home page

Web link 2.19: Online resource centre: **www.oxfordtextbooks.co.uk/orc/smith_smith/**

Web link 2.20: **http://msdn.microsoft.com/vbasic/**
Microsoft Visual Basic Developer Center

Web link 2.21: Online resource centre: **www.oxfordtextbooks.co.uk/orc/smith_smith/**

Web link 2.22: Online resource centre: **www.oxfordtextbooks.co.uk/orc/smith_smith/**

Web link 2.23: Online resource centre: **www.oxfordtextbooks.co.uk/orc/smith_smith/**

Web link 2.24: Online resource centre: **www.oxfordtextbooks.co.uk/orc/smith_smith/**

3   How to evaluate a model

Model evaluation is as crucial to modelling as the replication of measurements is to experimental work. Without model evaluation we have no idea of how different from reality the model simulation could be, we cannot be certain that the model behaves in the way we think it should, and we cannot determine which parts of the model are most likely to affect the result. We *can* use the model without evaluating it first, but we *should not*, just like we *can* use an unreplicated measurement to provide information about a system, but it is *better not to*.

Why a model needs to be evaluated – the last model did not work!

Your predecessor developed a model of soil carbon change for cultivated lands...

The model was used to estimate if cultivation could affect the climate.

The model predicted that cultivation would only have a small effect ... so nothing was done!

The model had not been evaluated, so was giving misleading results.

In fact, the effect was very significant!

Nothing was done, so soils continued to be degraded and carbon dioxide continued to be released.

Climate change escalated!

Mars was in trouble...

There are two reasons for evaluating a model: to check for human error in the construction of the model, and to check for circumstances where the model is likely to fail, due to the model being a simplification of reality. Scientific philosophy tells us that hypotheses can only be disproved—they cannot be proved; and the same applies to a model. All we can do is to check under what circumstances it is likely to fail; we cannot confirm that it *will* work. If a model has not been disproved, there is no reason to believe it will not work in similar conditions. If you do disprove the model, it is doubtful that the model will work. Modellers are always looking for the model to fail, even after a model has been constructed into an application. The failure of a model tells us something more about the system, and it is by this means that progress is made.

3.1 Decide what type of evaluation is needed

A thorough evaluation of a model will do the following three things.

1. It will determine the **accuracy** of the simulation. How confident can we be in the results?
2. It will analyse the **behaviour** of the model. Does the model respond in the expected way to changes in the conditions of the simulation?
3. It will resolve which **components** of the model are most important in determining the results.

However, a crucial prerequisite, before any quantitative model evaluation, is to assess the simulated results graphically. A **graphical analysis** takes a quick look at the results. Do they look approximately correct? What are the likely sources of error? It should be possible to spot by graphical analysis everything that is quantified in the model evaluation; the model evaluation should present no great surprises. A graphical analysis clarifies the interpretation of calculated statistics, helps to determine the questions that need to be asked in the evaluation, and saves time by not evaluating a model that is very obviously wrong. So, before you go any further, plot your results!

Having plotted the results, a model that appears to provide a reasonable fit to measured values should be evaluated quantitatively. The different types of model evaluation are groups of techniques that can be used to accomplish each of the above aims, namely, the determination of the accuracy, behaviour and important components of the model. It is not always necessary to use every form of analysis to evaluate the performance of a model. Which analyses should be used depends on the nature of the application of the model and so how important it is to be confident of each of the above aims.

A **quantitative analysis** determines the **accuracy** of the simulation, and is needed in almost all applications. It is a collection of statistical tests that allows the simulated values to be compared to measured values, and tells us how confident we can be in the simulated results. Some tests determine the coincidence between measured and simulated values, that is, do they have the same value and, if not, how different are they? Other tests determine the association between the measured and simulated values, that is, do they follow the same trend or do low and high measured and simulated values not occur at the same time?

A **sensitivity analysis** evaluates the **behaviour** of the model, and is used to determine whether the model responds in the expected way to changes in the conditions of the simulation. Model components that describe the conditions of the run (the input variables and fixed parameters) are adjusted so that the response of the model to changes in these components can be assessed. The output from the analysis is a plot of the changes in the simulated values against changes in the model components. The validity of the model response can be assessed in a qualitative way, from expert judgement (subjective sensitivity analysis), or in a quantitative way, by a comparison against experimental data. Any model that is to be used outside the range of conditions for which it was originally developed should undergo a thorough sensitivity analysis.

An **uncertainty analysis** resolves the importance of the model **components**. It is closely related to a sensitivity analysis, and indeed many of the same techniques and software packages are used. Whereas a sensitivity analysis defines how the model responds to changes in its components, an uncertainty analysis determines how much uncertainty is introduced into the model output by each component of the model. The values of the components of the model could be defined as input variables or fixed parameters. The uncertainty in the model components is used to determine a series of different starting values, and the simulation is run for each and for different combinations. The output from this process is the uncertainty in the results associated with the uncertainty in each component of the model. This information can be used to determine how much effort should be focused into improving measurements of the different input values; one application of an uncertainty analysis is to prioritise future experimental work. An uncertainty analysis might also be needed if the uncertainty in input components is likely to change. For instance, in a simulation using weather data, the quantitative analysis of the model might provide information about the accuracy of the simulations using weather data collected from a meteorological station on the site of the field experiment. Further applications may require simulations for sites where there is no meteorological station, as the nearest station might be 10, 20 or 50 km away. The uncertainty in the weather data increases with the distance of the meteorological station from the site. It is important, in this case, to understand how the uncertainty in the weather data introduces uncertainty in the model because the uncertainty in the weather data is likely to change.

Having decided what type of evaluation is needed in your specific application, you need more detail about the evaluation procedures to be able to thoroughly evaluate your model. In the following sections we describe each type of evaluation procedure in detail.

3.2 Plot the results (graphical analysis)

The type of plot needed and the information included in the plot depends on which of the three aspects of model evaluation is being considered, namely, the accuracy of the simulation, the behaviour of the model components, or the importance of model components. Common to all aspects of the evaluation is the guiding principle that the plot should give an immediate impression of how well the model is performing; if it does not do that, then it is no more than a pretty picture used to fill up a report, and can even give a misleading impression of good model performance. In this section we discuss some potential pitfalls of a graphical analysis of model performance. A graphical analysis should always be an integral part of the model evaluation. The first step in the model evaluation is to plot the things that are being evaluated, the second is to quantify the evaluation, and the third is to refer back to the plot and check that the graphical and statistical analyses provide the same information. This section on graphical analysis is discussed separately merely to emphasise its importance and to avoid the importance of graphical analysis becoming lost in discussions of statistics.

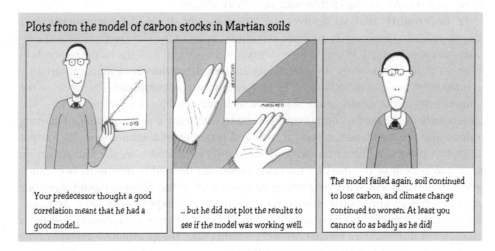

Plots from the model of carbon stocks in Martian soils

Your predecessor thought a good correlation meant that he had a good model...

... but he did not plot the results to see if the model was working well.

The model failed again, soil continued to lose carbon, and climate change continued to worsen. At least you cannot do as badly as he did!

3.2.1 Plots to reveal the accuracy of the simulation

A plot to reveal the accuracy of a simulation should include the measured and simulated values of the result that is needed from the model, not a precursor to, or some surrogate of, the needed result. For instance, if you are interested in predicting damage to Martian fern from attack by the Martian fern weevil, a plot of simulated and measured weevil numbers against time does not directly reveal whether the model is accurately predicting damage to the fern; a plot of measured and simulated fern damage is needed to reveal this. The plot of weevil numbers will be useful in determining why the model is failing, but that is plotted at a later stage of the evaluation, and is accompanied by a clear hypothesis or research question as to the source of the model error.

   Usually, the measured and simulated values should be presented on the same plot. This allows differences to be clearly highlighted and avoids any erroneous impression of the relative size of simulated and measured values that can arise due to changes in the axes used to plot the results. There are two ways in which results are commonly presented, namely, simulated values plotted against measured values, or simulated and measured values plotted against some other variable that is used by the model.

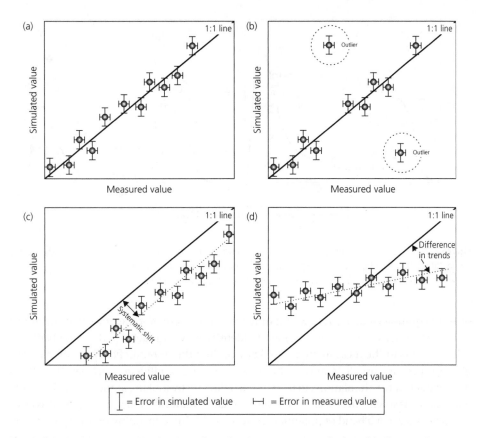

Figure 3.1 The features revealed by plots of simulated against measured values: (a) all points show a good fit; (b) outliers; (c) a negative bias in simulations; (d) a difference in trends of measured and simulated data.

Simulated values can be plotted on the y-axis against measured values on the x-axis (Fig. 3.1). The accuracy of the simulation is revealed by the proximity of the points to the 1:1 line (Fig. 3.1(a)). Any outliers will be highlighted by this type of plot (Fig. 3.1(b)), any systematic shift in the simulated values with respect to measured values will be revealed (Fig. 3.1(c)), and any differences in the trends in simulated and measured values will also become apparent (Fig. 3.1(d)).

Alternatively, simulated and measured values can both be plotted on the y-axis, against some other variable on the x-axis (Fig. 3.2). The variable plotted on the x-axis is usually an input to the model that can be used to specify the conditions of the simulation, such as time or weather data. This type of plot is not as good as the simulated versus measured plot at highlighting outliers, systematic shifts or differences in trends, but allows patterns in errors to be identified. For instance, the simulated values in Fig. 3.2(a) show a clear forward shift in time with respect to the measurements, indicating that the rate of some process in the model is too slow. The simulated values in Fig. 3.2(b) show a downward shift in value with respect to the measurements, while replicating the timing

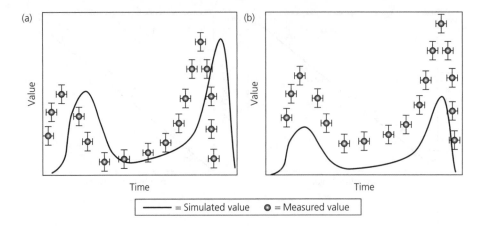

Figure 3.2 An Illustration of how plots of measured and simulated values against an input variable to the model (time) can reveal patterns in errors: (a) simulated values showing a forward shift in time; (b) simulated values showing an upward shift with respect to measured values.

of events quite accurately. This would indicate that the size of some process in the model is inaccurate. This type of plot can help to identify why a model is performing badly.

If the errors in the measurements are known, then these should be represented on the plot as error bars, usually showing the standard error or confidence interval; this allows the accuracy of the simulations to be estimated directly by looking at the graph. Similarly, if the model is stochastic, variations in the simulated values should be represented as error bars (Fig. 3.3(a)) or a band of potential results (Fig. 3.3(b)), depending on the type of plot.

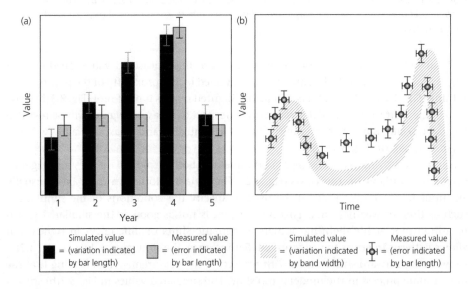

Figure 3.3 The representation of errors in measured and simulated values, allowing significant differences to be discerned by eye: (a) error bars; (b) a band of potential.

The scales of the axes should be selected so that it is possible to distinguish differences as small as the acceptable error. An example of how an apparently sensible choice of axis scale can give a misleading impression of good model performance comes from the field of dynamic carbon modelling (as used on Earth or on Mars to detect changes in soil carbon under different management; see Chapters 2 and 4). Dynamic carbon models are often initialised with the measurement of the carbon content of the soil. The initialised model is used to run forward in time, predicting the changes in soil carbon content with changes in management. Soil carbon contents vary by orders of magnitude between mineral and organic soils (like peats). Therefore, simulated values are often plotted against measured values using logarithmic scales, so that comparisons for mineral and organic soils can be included on the same plot. If the acceptable error is taken to be $2 \, t\,C\,ha^{-1}$, then this error can easily be seen for soils with a low carbon content, but the logarithmic scale makes it difficult to distinguish errors of this size in highly organic soils that might have carbon contents at the top end of the scale shown in Fig. 3.4(a). This gives a false impression of good model performance in highly organic soils. In this case, it would be more informative if the results for the different types of soils were plotted on different graphs, with a linear scale and the bottom of the scale selected to allow the acceptable error to be distinguished. To allow a comparison of the errors, as shown in Fig. 3.4(b–d), the range of values included on the axes of the different plots (i.e., the top value minus the bottom value) should be the same.

Finally, plots to reveal the accuracy of simulations should avoid presenting the results in a confusing or complicated manner. When working with complicated models, it is often useful to investigate more than one factor at a time because these multiple factors may interact with each other to produce a different result. For instance, if modelling the reduction in slug feeding due to the predation of nematodes on slugs, then the concentration of nematodes to slugs influences the effect on feeding reduction of the time of exposure. It is tempting to plot these three factors, namely, the percentage reduction in feeding, the time of exposure and the concentration of nematodes, on a three-dimensional graph (Fig. 3.5(a)). However, it is difficult to obtain information from such a plot. More information might be provided by a graph that omits some of the results available, but presents fewer results in a stepped fashion on a two-dimensional plot (Fig. 3.5(b)). This allows results to be more easily read from the graph, and a comparison of measured and simulated values to be made by eye.

Plots to reveal the accuracy of the simulation should

1. show the result that is needed from the model,
2. include simulated and measured values on the same plot,
3. show errors in measurements as error bars,
4. show the variation in simulations as error bars or as a band of potential results,
5. use axes that highlight the acceptable error in the simulation, and
6. avoid confusing and complicated presentation.

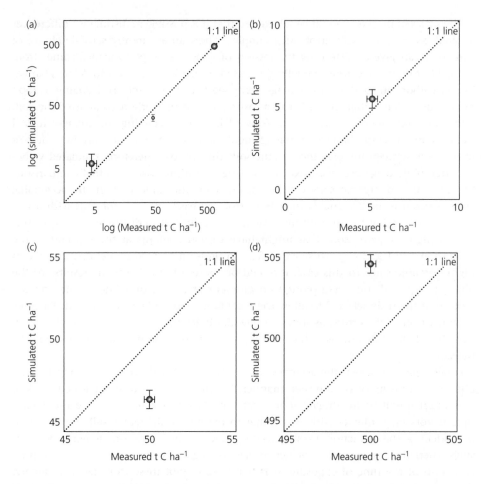

Figure 3.4 An Illustration of how the choice of scale can affect the impression of good model performance. (a) A logarithmic scale, selected to allow a range of results to be plotted on the same graph, obscures the errors in the points at higher values. (b–d) The same information plotted on a linear scale reveals the greater error in the higher values.

3.2.2 **Plots to illustrate the behaviour of the model components**

A plot to illustrate how changes in the model components affect the behaviour of the model usually requires the result that is needed from the model to be plotted against the change in the model component. The change in the model component can usually be defined quantitatively using the parameter or input variable value that is being adjusted during the sensitivity analysis. Plotting these results is an integral part of the sensitivity analysis, and cannot be done in advance.

Graphical plots to illustrate the behaviour of the model components are prone to over-complication. Different components of a model will often interact with each other, so it is important to investigate the effects of simultaneous changes in many different components of the model. However, a plot is only of value if it can be interpreted by eye.

(a)

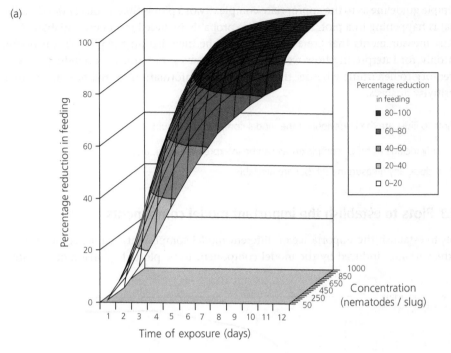

(b)

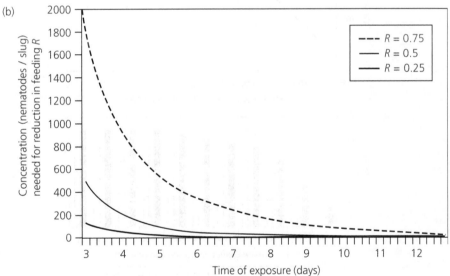

Figure 3.5 An illustration of how multidimensional plots can reduce the clarity of presentation. (a) A three-dimensional plot of percentage reduction in slug feeding with time of exposure to nematodes and concentration of nematodes to slugs. (b) A two-dimensional plot of the same information (R = percentage reduction in feeding/100).

A simple guideline as to the appropriate complexity of a plot is that if you cannot describe what is happening in a plot, then you have probably included too many variables.

Any measurements that are available should be included on the plot. This is useful, not only for interpreting how well the model behaves, but also for providing a visual inventory to determine whether there is sufficient information for the full quantitative sensitivity analysis.

Plots to illustrate the behaviour of the model components should

1. only include as much complexity as can be interpreted by eye, and

2. include any measurements that are available.

3.2.3 Plots to establish the important model components

Plots to establish the importance of different model components require some measure of the variation induced by the model component to be plotted against a quantitative

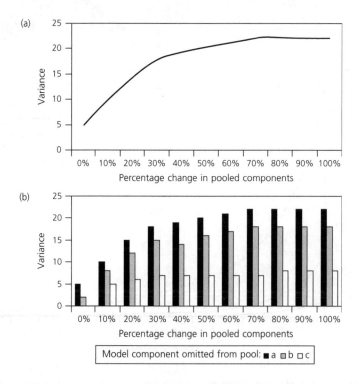

Figure 3.6 Different ways of presenting the results of an uncertainty analysis. (a) A plot of the change in variance against a pooled change in the model components. (b) A bar chart showing the change in variance against a pooled change in the model components, with one component omitted from the pool. (c) A radar plot showing the change in variance at a given change in variance (the shape of the area acts as a 'thumbprint' of the results). (d) As in (c), with the order of the plotting components e and h swapped (illustrates the need for caution in interpreting these plots).

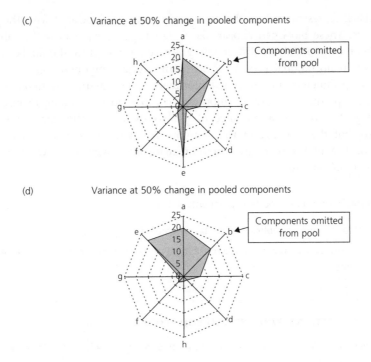

Figure 3.6 (*continued*)

measure of the change in the model component. A quantitative measure of the change in the model component is given by the value of the parameter or input variable that is adjusted to change the influence of the model component. A measure of the variation is given by the variance, that is, the square of the sum of the deviations from the mean divided by the number of replicates less one. The replicates of the simulations can be derived from a **Monte Carlo simulation**, performed as part of the model uncertainty analysis. Again, plotting these results is an integral part of the uncertainty analysis and cannot be done in advance.

As this type of analysis often requires simultaneous changes in many different components of the model, plots can become very complex and difficult to interpret. As discussed in the previous section, only as much complexity should be included in each plot as can be interpreted by eye.

A number of different approaches are commonly used to allow such complex results to be presented. Changes in the model components may be pooled, so that the same percentage change is incorporated in all of the related components. The results can be plotted as the variance against the percentage change in only one or two pools or components, allowing graphs to be two- or three-dimensional (Fig. 3.6(a)). The importance of each component can be analysed further by omitting one component at a time from the pool of components. The results can be plotted as bar charts, showing the variance associated with each component (Fig. 3.6(b)). Polar charts, showing the variance for multiple components, are plotted around a central point, and can provide a more visual representation of the changes in variance associated with the different components. These plots can appear like a signature, showing how the influence of the different components is distributed. However, these plots should be used with care if there is no rational reason for positioning one component beside another, as the order of parameters chosen for the plot determines whether the signature appears like an arrow pointing toward an important component (Fig. 3.6(c)) or just a sloped surface (Fig. 3.6(d)). The importance of the component is the same in Figs 3.6(c) and 3.6(d), but the ordering of the components in the plot gives a very different impression. The plot should give an immediate impression of how important the different components are.

Plots to establish important model components should

1. only include as much complexity as can be interpreted by eye, and

2. use measures such as changes in the distribution of results to reflect the importance of components.

SELF-CHECK QUESTIONS: WHAT WOULD YOU PLOT?

1. You develop a model that produces a series of estimates of soil carbon at a single site over one hundred years.

   Q: Which of the following would be useful to plot to see how well the model is working?
   a. Measured values versus model estimates
   b. Model estimates against time with measurements on the same graph
   c. Model estimates against map coordinates of the site
   d. Model estimates against log of time with measurements on the same graph

   [A: a and b are useful. Plot a shows how closely the measured and modelled values compare and how close they are to the 1:1 line (perfect agreement). Plot b shows how well the model captures the dynamics of the change in soil carbon over time. Plot c is not appropriate as the model is for one site only. For plot d, there is no reason to plot against log of time.]

2. You are interested in predicting the damage to Martian fern from attack by the Martian fern weevil. You develop a model to do this that has various outputs.

   Q: Which plot would help you to determine if the model is working?
   a. Simulated and measured weevil numbers against time
   b. Measured and simulated fern damage

   [A: Plot b is the critical plot. Plot a might be useful in determining how the model is failing (if it is), but is used to interpret why the model is not working rather than to confirm that it is, or is not, working well.]

3. Q: Why is it useful to show the error bars of the measurements on a plot showing simulated and measured values on the same graph?
   a. It shows that the measurements are not very good
   b. It shows whether or not the model is within measurement error
   c. It shows that the model is not very good

   [A: b.]

3.3 Calculate the accuracy of the simulation (quantitative analysis)

A quantitative analysis of a model simulation tells us how well the simulated values match measured data. Two types of analysis are most frequently used, namely, an analysis of coincidence, and an analysis of association, and a thorough analysis will include both.

An analysis of coincidence tells us how different the simulated and measured values are. Standard formulae are used to calculate statistics that quantify the difference between simulated and measured data. These statistics can express this as the total difference, can separate out positive or negative bias, or can exclude any difference associated with variations in the measurements. Each of these ways of expressing difference provides unique information.

An analysis of association tells us how well trends in measured values are simulated. It is important to analyse the trends as well as the difference between simulated and measured values because this tells us if the model is showing the right responses to changes in conditions, and so whether it is likely to continue to simulate the system well under different conditions. A model with a small difference as well as a high association between simulated and measured values (as illustrated in Fig. 3.7(a)) is simulating the measured data accurately. It is, however, possible for the difference between simulated and measured values to be small (a good model), but the association to be low (a bad model), as illustrated in Fig. 3.7(b); this might suggest that the fit between simulated and measured data is no more than good luck! Equally, as illustrated in Fig. 3.7(c), it is possible for the measured and simulated values to be closely associated (a good model), but the difference between them to be high (a bad model); this might suggest that

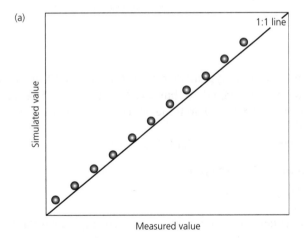

Figure 3.7 An Illustration of coincidence and association between measured and simulated values: (a) high coincidence and association (good model); (b) high coincidence but low association (bad model); (c) high association but low coincidence (bad model).

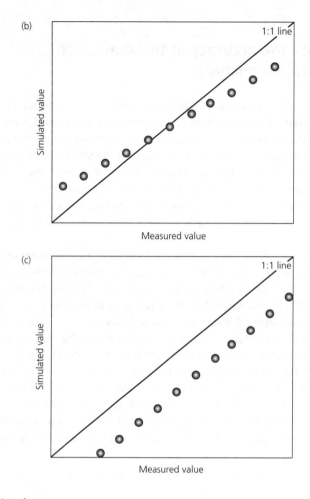

Figure 3.7 (*continued*)

the model is good, but some systematic error is causing a shift in the simulations and preventing a good fit. What you do to improve the model is different if the association is high and the coincidence low, than if the association is low and the coincidence high.

A thorough quantitative analysis should include both

1. an analysis of coincidence (difference), and
2. an analysis of association (trends).

The most crucial (and often limiting) component of a quantitative analysis is the measured data against which simulations are compared. The nature of the available data dictates which statistics are most appropriate, and determines how much information can be obtained about the model. The accuracy of the measurements limits the accuracy of the model evaluation. A quantitative analysis can only tell us how well the model simulates the data it is compared against; it cannot confirm that the model will work well in other situations, but can only confirm that the model does not work in this situation. We can, however, say that, if a model simulates measurements under specific conditions accurately, then, in all probability, it will simulate measurements under similar conditions with equal accuracy. If it does not, the question as to why it does not is of great interest to the model developer (what has changed to stop the model working?), to the experimentalist (is there some error in the measurements?) and to the scientist (is there some unknown process that has not been accounted for?). For these reasons, model evaluation should never be considered to be complete; further evaluation will always provide more information about the model and the system—if you know how to use it.

A model can only be properly evaluated against independent data, that is, data that was not used to develop the model. The procedures of model development, involving the derivation of equations, parameter fitting or other data-dependent methodologies, amalgamate the effects of any processes that have not been included in the model into the description of those processes that are included. If the model is then evaluated using the same data, any errors introduced by amalgamating the description of these processes will not be exposed. If the model is evaluated against independent data, then it is likely that the effect of the missing process will be different with respect to the process it has been amalgamated into, so the model evaluation will show an error, thereby exposing this fault. It is easy to fit parameters to force a set of equations to accurately simulate any measured data; a useful model should be able to simulate independent data with some degree of accuracy as well.

A quantitative analysis should compare simulated results to **independent** measured data.

A quantitative analysis of model performance should use independent measurements for the full range of conditions for which the model is to be used, and should provide assessments of both association and coincidence between simulations and measured data. A collection of statistics used by modellers to evaluate their models, and a scheme for deciding which statistics to use depending on the nature of the measured data available, are given in the following sections.

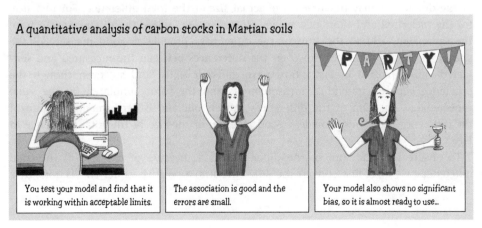

A quantitative analysis of carbon stocks in Martian soils

You test your model and find that it is working within acceptable limits.

The association is good and the errors are small.

Your model also shows no significant bias, so it is almost ready to use...

3.3.1 Analysis of coincidence

Total difference

The lack of coincidence can be quantified as the **difference** between simulated and measured values. The difference can be calculated for each simulated value and its corresponding measurement, simply by subtraction, as follows:

$$\text{Difference} = O - P$$

where O is the measured (observed) value and P is the simulated (predicted) value.

The total difference in the simulation from the measurements is the sum of all the differences at all measurements, namely,

$$\text{Total difference} = \sum_{i=1}^{n} (O_i - P_i)$$

where O_i is the ith measured value, P_i is the ith simulated value, and n is the total number of measured and simulated pairs being compared.

Reminder of notation

Note that the Greek notation $\sum$ is used to denote 'sum of', the subscript $\sum_{i=1}$ to denote 'from $i = 1$', and the superscript $\sum^{n}$ to denote 'to n'. So $\sum_{i=1}^{n}$ means the sum of the series from elements $i = 1$ to n. If, for example, n is 2, then the total difference is

$$\sum_{i=1}^{n} (O_i - P_i) = (O_1 - P_1) + (O_2 - P_2).$$

However, because the difference between pairs of simulated and measured values is sometimes positive (the measured value is greater than the simulated value) and sometimes negative (the measured value is less than the simulated value), the sum of the differences may not reflect the actual *size* of the total difference. For instance, for the measured and simulated values plotted in Fig. 3.8(a), the differences all have the same sign, and so the sum of the differences provides a good measure of the size of the total difference. However, the differences between the measured and simulated values plotted in Fig. 3.8(b) have different signs, and so, even though the difference between individual measurements and the value simulated for the same time is large, the sum of the differences cancels out and is very small. The sum of the differences does not provide a good measure of the size of the total difference in Fig. 3.8(b).

To ensure that the sum of the differences is a measure of the size of the total differences, we must first convert the difference into a positive value. This is done by a

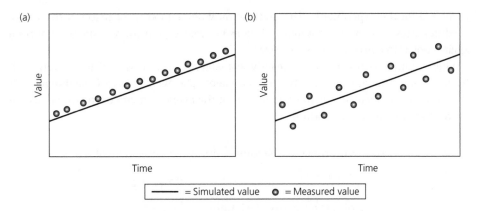

Figure 3.8 An illustration of two types of error. (a) The differences between the simulated and measured values all have the same sign; the sum reflects the size of the total difference. (b) The differences between the simulated and measured values have different signs; the sum does not reflect the size of the total difference.

simple mathematical trick, namely, taking the square of the number and then the square root as follows:

$$\text{Total size of the difference} = \sum_{i=1}^{n} \sqrt{(O_i - P_i)^2}.$$

The total size of the difference is dependent on the number of measurements included; the greater the number of measurements, the greater the total size of the difference will be. To get a statistic that is not dependent on the number of measurements, the average is taken, by dividing by the total number of measurements. The average size of the difference is given by

$$\text{Average size of the difference} = \frac{\sum_{i=1}^{n} \sqrt{(O_i - P_i)^2}}{n}.$$

Before we had computers, this was not an easy value to calculate; try calculating the square root of 29.65 in your head—it is not easy! So the average size of the difference was most often defined by the square term as follows:

$$\text{Average square of the size of the difference} = \frac{\sum_{i=1}^{n} (O_i - P_i)^2}{n}.$$

The important thing about this statistic, is that the *order* of the average square values calculated for a set of experiments will be the same as the *order* of the average values calculated for the experiments. This statistic provides a good comparison of the size of the differences in a set of experiments; it does not matter that the statistic has not been square rooted to convert it back to a size.

When computers became more readily available, scientists wanted to convert their results back to the same units as the measurement. The average size of the difference is now commonly calculated as the square root of the average square of the difference, or the **root mean squared deviation**.

Total difference in the same units as the measurement

$$\text{Root mean squared deviation} = \sqrt{\dfrac{\sum\limits_{i=1}^{n}(O_i - P_i)^2}{n}},$$

where O_i is the ith measured value, P_i is the ith simulated value, and n is the total number of values being compared.

The root mean squared deviation is not, of course, identical to the average size of the difference (see below), but, as for the square value, the important thing about this statistic is that the *order* of the values will be the same, so it provides an accurate comparison of the size of the differences in a set of experiments.

Example illustrating why root mean squared deviation is not identical to the average size of the difference . . .

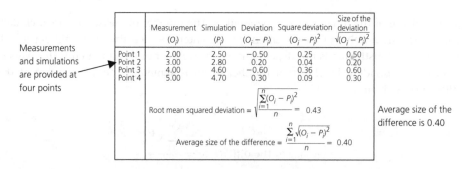

		Measurement (O_i)	Simulation (P_i)	Deviation $(O_i - P_i)$	Square deviation $(O_i - P_i)^2$	Size of the deviation $\sqrt{(O_i - P_i)^2}$
Measurements and simulations are provided at four points	Point 1	2.00	2.50	−0.50	0.25	0.50
	Point 2	3.00	2.80	0.20	0.04	0.20
	Point 3	4.00	4.60	−0.60	0.36	0.60
	Point 4	5.00	4.70	0.30	0.09	0.30

$$\text{Root mean squared deviation} = \sqrt{\dfrac{\sum\limits_{i=1}^{n}(O_i - P_i)^2}{n}} = 0.43$$

$$\text{Average size of the difference} = \dfrac{\sum\limits_{i=1}^{n}\sqrt{(O_i - P_i)^2}}{n} = 0.40$$

Average size of the difference is 0.40

The root mean squared deviation expresses the total difference in the same units as used for the measurement. In the example of changes in carbon stocks in the cultivated lands of Mars, carbon stocks were measured in $t\,ha^{-1}$, and so the value of the root mean squared deviation will also be given in $t\,ha^{-1}$. This can be useful, as the scientist will usually have some idea of what size of difference from the measurement is acceptable (RMS_{acc}), and so can state whether the simulation is sufficiently accurate or not.

Loague and Green (1991) suggested another way of determining whether the total difference is acceptable, namely, by re-expressing the total difference as a percentage. This statistic, known as the root mean squared error, allows the acceptability of the model to be decided by comparing to a standard value for the percentage acceptable error ($RMSE_{acc}$). For example, an error of 10% total difference may be considered to be acceptable for a particular purpose, so any value of the root mean squared error of less than 10% says that the model is sufficiently accurate.

Total difference as a percentage

$$\text{Root mean squared error} = \frac{100}{\overline{O}} \times \sqrt{\frac{\sum_{i=1}^{n}(O_i - P_i)^2}{n}},$$

where $\overline{O}$ is the average of all the measurements, O_i is the ith measured value, P_i is the ith simulated value, and n is the total number of values being compared.

The root mean squared error allows a comparison with a standard percentage. However, the percentage error considered to be acceptable is still an arbitrary choice. Another approach, described by Smith *et al.* (1997), avoids the arbitrary choice of acceptable error by comparing the statistic to the value that would be obtained if the simulated values were at the 95% confidence interval from the measurement. For an excellent, full explanation of confidence intervals, see Townend (2002).

The 95% confidence interval is the range around the mean value within which 95% of the population is expected to be found. Each measurement that the simulated values are compared to is actually the mean of a sample of replicate measurements, a replicate being a repeat of the measurement taken at the same time and under identical experimental conditions (see Fig. 3.9). If the simulation is within the 95% confidence interval, then it is as good an estimate of the mean measured value as the replicate measurements themselves; there is no sense in trying to make the simulation a more accurate estimate of the mean value than this. The value of the statistic that would be obtained if the simulated values were at the 95% confidence interval from the measurement represents the best-case evaluation for the simulation against this measured data.

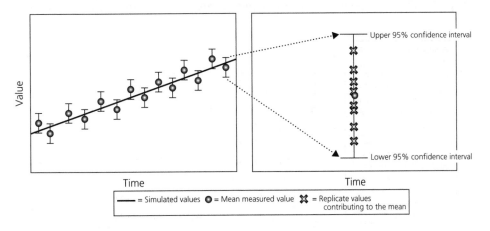

Figure 3.9 An illustration of how replicate measurements relate to the mean measured values and the 95% confidence interval.

If the measurements have been replicated, the distribution of the replicated values allows the mean measurement at each point and the standard error in the measurements to be determined.

The mean measurement is given as follows:

$$\text{Mean value of the } i\text{th measurement}, O_i = \frac{\sum_{ij=1}^{m} O_{ij}}{m},$$

where O_{ij} is the jth replicate of the ith measurement, and m is the total number of replicates of the ith measurement.

The standard error is given as follows:

$$\text{Standard error}, SE_i = \frac{\sum_{ij=1}^{m} (O_i - O_{ij})^2}{\sqrt{m} \times (m - 1)}.$$

The 95% confidence interval can be obtained from the standard error multiplied by the Student's t at 95% probability. The Student's t is a statistic that allows us to determine the probability that two samples of replicates are from the same population. The Student's t is defined as the difference in the means of each sample of replicates divided by the standard error of the samples, that is,

$$\text{Student's } t = \frac{\sqrt{(O_{i,1} - O_{i,2})^2}}{SE_i},$$

where $O_{i,1}$ is the mean measurement of the sample 1 replicates and $O_{i,2}$ is the mean measurement of the sample 2 replicates.

The Student's t is defined in this way because, for a normally-distributed population, as the probability that the two samples of replicates are different from the mean increases,

P-values: multiply by 100 to give the percentage probability that the null hypothesis (the two samples are not from the same population) can be rejected.

df: degrees of freedom, equal to the number of replicates less the number of alternative outcomes (2), that is, $df = m - 2$.

P	0.10	0.05	0.025	0.01	0.005	0.001	0.0005
df = 1	3.076	6.314	12.706	31.821	63.657	318.31	636.62
2	1.886	2.920	4.303	6.965	9.925	22.326	31.598
3	1.638	2.353	3.182	4.541	5.841	10.213	12.924
4	1.533	2.132	2.776	3.747	4.604	7.173	8.610
...							
98	1.290	1.661	1.984	2.365	2.627	3.176	3.393
∞	1.282	1.645	1.960	2.326	2.576	3.090	3.291

Figure 3.10 A typical layout of a table of the Student's t distribution.

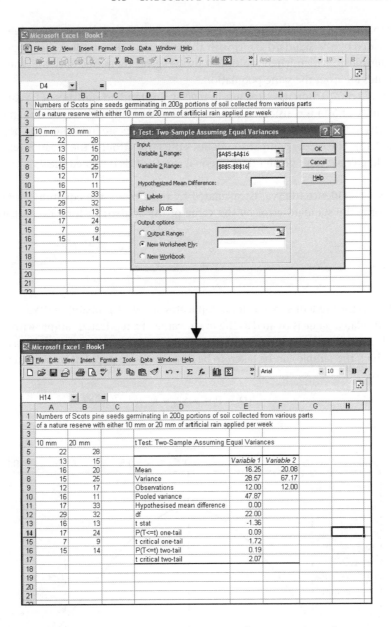

Figure 3.11 An example of the use the Analysis ToolPak to perform a *t*-test in Excel.

this ratio also increases. The values of the Student's *t* change in a standard way with the number of replicates and the probability that a value is different from the mean. Standard values of the Student's t have been determined by statisticians and can be obtained from look-up tables in statistical textbooks (see Fig. 3.10 and the 'Further reading' section, for example Chatfield (1983)) or from general spreadsheets (such as Microsoft Excel, see Fig. 3.11).

A comparison of the calculated value of t (the difference between the value and the mean divided by the standard error) with the standard t distribution allows the P value to be determined, which when multiplied by 100 gives the probability that the samples of replicates are from the same population.

The model comparison uses this test in reverse. In standard statistics, a P value of 0.05 (a probability of 5%) is considered to be a significant result; a P value less than or equal to 0.05 tells us there is a low probability that the replicates are from the same population (see Fig. 3.12). The model comparison defines the range around the mean measurement that would give a P value greater than or equal to 0.05; this is termed the 95% confidence interval.

The 95% confidence interval is calculated by rearranging the equation for Student's t to multiply the standard error by the value of the Student's t for 95% probability as follows:

$$\sqrt{(O_i - O_{ij})^2} = SE_i \times t_{m,95},$$

where $t_{m,95}$ is the Student's t value for m replicates and 95% probability (P value of 0.95).

If the simulated value is treated as the mean value of a replicated sample with a similar standard error to the measurements, then the range around the mean measurement within which the sample can be taken to be from the same population can be calculated as follows:

$$\sqrt{(O_i - P_i)^2} = SE_i \times t_{m,95}.$$

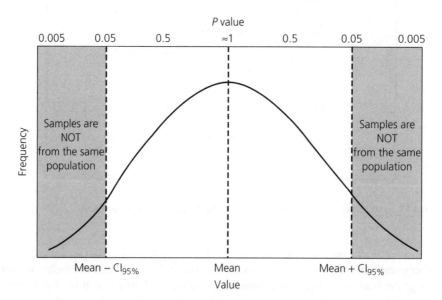

Figure 3.12 An illustration of how the difference in the mean values can be used to distinguish samples that are not from the same population. Here Cl$_{95\%}$ denotes the 95% confidence internal.

Substituting this into the equations for the total error gives the 95% confidence interval in the root mean squared deviation (RMS_{95}) and in the root mean squared error ($RMSE_{95}$). This is the range within which the simulation can be considered to be equivalent to the measured mean.

Significance of the total difference

The 95% confidence interval

in the root mean squared deviation, $$RMS_{95} = \sqrt{\frac{\sum_{i=1}^{n} (SE_i \times t_{m,95})^2}{n}}$$

and in the root mean squared error, $$RMSE_{95} = \frac{100}{\overline{O}} \sqrt{\frac{\sum_{i=1}^{n} (SE_i \times t_{m,95})^2}{n}},$$

where $\overline{O}$ is the average of all the measurements, SE_i is the standard error in the ith measurement, $t_{m,95}$ is the Student's t value for m replicates and 95% probability (P value of 0.95), and n is the number of measurements.

If the value of the statistic calculated from the measured and simulated values is greater than its 95% confidence interval, then the model could be improved. If the value is less than its 95% confidence interval, then the model is performing to within the accuracy of the measurements and cannot be improved further using this data.

If $RMS < RMS_{95}$, the model cannot be improved using this data.
If $RMSE < RMSE_{95}$, the model cannot be improved using this data.

The form of the equations allows the significance of the total error to be calculated using quoted standard errors, without the need to refer back to the original replicated measurements. Replicated measurements are not often provided in journal articles, but standard errors are usually quoted. Therefore, this method of evaluating the coincidence between the measured and simulated values is invaluable if, as is often the case, the evaluation relies on measurements from the literature.

Bias

The bias in the difference between the simulated and measured values can be calculated as the sum of the differences from the mean without forcing the value to be positive (by taking the square root of the squared difference, see Fig. 3.8), that is,

$$\text{Total bias in the difference} = \sum_{i=1}^{n} (O_i - P_i).$$

As for the total difference, this expression of the total bias will be greater as the number of measurements that are included in the comparison is increased. To obtain

a statistic that is independent of the number of measurements, the average is calculated by dividing by the total number of measurements. Addiscott and Whitmore
(1987) further re-expressed the bias as a percentage so that the result could be evaluated by a comparison to a standard acceptable percentage error. If measurements
are replicated, this statistic, known as the relative error, E, can also be evaluated
against the value of the statistic at the 95% confidence interval, in the same way as
described above.

Bias in the difference if measurements are replicated

$$\text{Relative error, } E \quad = \frac{100}{\overline{O}} \times \frac{\sum\limits_{i=1}^{n} (O_i - P_i)}{n}.$$

Significance of bias if measurements are replicated

$$\text{95\% confidence interval of relative error, } E_{95} = \frac{100}{\overline{O}} \frac{\sum\limits_{i=1}^{n} (SE_i \times t_{m,95})}{n}.$$

Here $\overline{O}$ is the average of all the measurements, O_i is the ith measured value, P_i
is the ith simulated value, n is the total number of values being compared, SE_i is
the standard error in the ith measurement, and $t_{m,95}$ is the Student's t value for m
replicates and 95% probability (P value of 0.95).

If the total error is significant, but the bias is insignificant, this suggests that, although
the accuracy of the model is low, the processes have been included in the right way.
If both the total error and the bias in the model are significant, this suggests that the
error is due to the model overestimating or underestimating some process in the system,
and major structural changes to the model may be needed before the accuracy of the
simulations can be improved.

If $E < E_{95}$, the model bias cannot be reduced using these data.

Due to the high cost of field experimentation, measurements, especially from older
long-term field trials, are often not replicated, and so standard errors are not quoted. This
would preclude model evaluation by a comparison with the 95% confidence interval.
In many areas of ecological and environmental science, changes in the system can be
very slow, and so long-term experiments provide information that cannot be obtained
from more modern experiments. It is very important to be able to evaluate the bias
in the model even if measurements are not replicated. Addiscott and Whitmore (1987)
provide a method that uses the variation across the different measurements (rather than
the replicates) to calculate the value of the Student's t. The bias is expressed as the mean
difference, M.

Bias in the difference if measurements are unreplicated

$$\text{Mean difference}, M = \frac{\sum\limits_{i=1}^{n} (O_i - P_i)}{n},$$

where O_i is the ith measured value, P_i is the ith simulated value, and n is the total number of values being compared.

The value of the Student's t can be calculated directly from the mean difference using the standard statistical formula for t.

Significance of the bias in the difference if measurements are unreplicated

$$\text{Student's } t = \frac{M \times \sqrt{n}}{\sqrt{\sum\limits_{i=1}^{n} \left(O_i - P_i - \left(\sum\limits_{i=1}^{n} (O_i - P_i)/n \right) \right)^2 \Big/ (n-1)}},$$

where O_i is the ith measured value, P_i is the ith simulated value, and n is the total number of values being compared.

The calculated value of the Student's t for M can then be compared to the t distributions available from statistical tables or from standard statistical packages (such as Microsoft Excel) to obtain the probability that the mean difference is statistically significant. A P value less than or equal to 0.05 suggests that the simulated values are showing significant bias with respect to the measured values. Note that this calculation of significance uses the variation in the difference over all the measurements, rather than the variation in replicates at each individual measurement point, and so can give misleading results if the bias varies greatly at different measurements. If replicates are available, it is preferable to calculate bias using the relative error.

Difference excluding variations in the measurements

Whitmore (1991) provides a statistic referred to as the lack of fit, LOFIT, which separates out the errors due to variations in the measurements from the errors due to the difference between the simulated and the measured values. Only errors due to the difference between the simulated and measured values (the LOFIT) can be attributed to failure of the model, so it makes sense to use only these errors when evaluating model performance. The significance of LOFIT can be determined using an F test to compare the variances of LOFIT and the error associated with variations in the measurements. These equations can appear complicated, but in fact only include replicate measurements and simulated values. Careful substitution of the appropriate values that are output from the model into the formulae below allows a thorough model evaluation to be completed quickly. A spreadsheet in Microsoft Excel that calculates these statistics for you is included on the website (**Web link 3.1**).

The probability associated with the F value can be obtained by a comparison with standard statistical tables of the F distribution. Alternatively, $\sqrt{F}$ can be compared to the distribution of the Student's t.

Difference between measured and simulated values excluding error associated with variations in measurement

$$\text{LOFIT} = \sum_{i=1}^{n} m_i(O_i - P_i)^2.$$

Significance of difference between measured and simulated values excluding error associated with variations in measurement

$$F = \frac{\sum_{i=1}^{n}(m_i - 1) \times \text{LOFIT}}{n\sum_{i=1}^{n}\sum_{j=1}^{m_i}((O_{ij} - P_i) - (O_i - P_i))^2}.$$

Here, m_i is the number of replicates of the ith measurement, O_i is the mean value of the ith measurement, P_i is the ith simulated value, n is the number of simulated and measured pairs being compared, and O_{ij} is the jth replicate of the ith measurement.

Of the statistics considered here, LOFIT provides the most thorough analysis of coincidence, and so is the best single statistic to use to define coincidence. However, actual values of replicated measurements are needed to calculate LOFIT, not just the standard error around the mean. If replicates are not available, as is often the case in measurements obtained from the literature, a combination of the statistics already described to calculate total error and bias will provide an adequate assessment. A decision tree is given in Fig. 3.13 that should help to guide you through the choice of statistics for an analysis of coincidence.

There are many other measures of coincidence, such as **modelling efficiency, coefficient of determination** and **coefficient of residual mass** (Loague and Green, 1991). The statistics described here should provide a sufficiently comprehensive evaluation of coincidence for most purposes.

3.3.2 **Analysis of association**

The degree of association between the measured and simulated values can be determined using the standard statistic, the **sample correlation coefficient** (see the 'Further reading' section, Townend, 2002, for an excellent, full explanation; see Chatfield, 1983, for the statistical proof).

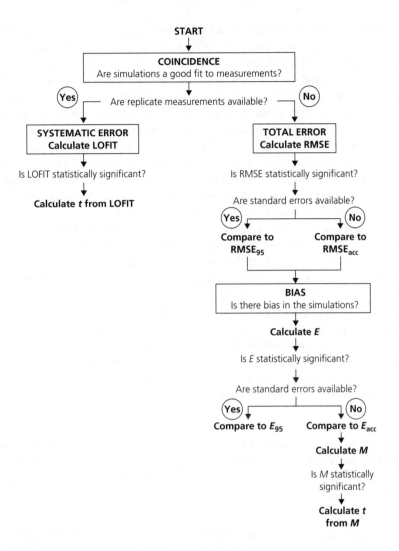

Figure 3.13 The selection of statistics for an analysis of coincidence.

Degree of association between measured and simulated values

Sample correlation coefficient, $r = \dfrac{\sum\limits_{i=1}^{n} (O_i - \overline{O})\,(P_i - \overline{P})}{\sqrt{\sum\limits_{i=1}^{n} (O_i - \overline{O})^2}\ \sqrt{\sum\limits_{i=1}^{n} (P_i - \overline{P})^2}},$

where O_i is the ith measured value, $\overline{O}$ is the average measured value, P_i is the ith simulated value, and $\overline{P}$ is the average simulated value.

The sample correlation coefficient can have any value from -1 to 1. A correlation coefficient of 1 indicates a perfect positive correlation, that is, the simulated values are strongly associated with the measured values, suggesting that the model is performing well. A correlation coefficient of 1 could be obtained from any of the sets of simulated and measured values plotted in Fig. 3.14(a). The correlation coefficient calculated in this way gives no information about how well the measured and simulated values match each other; it only tells us how well the *trends* are simulated.

A sample correlation coefficient of -1 indicates a perfect negative correlation, that is, the simulated values are strongly associated with the measured values, but there is something wrong with the way the model is performing and the results are coming out the wrong way round. A correlation coefficient of -1 could be obtained from any of the sets of simulated and measured values plotted in Fig. 3.14(b).

A sample correlation coefficient of 0 (Fig. 3.14(c)) indicates no correlation between the simulated and measured values. The model is performing badly, and, even if the analysis of coincidence indicates a good fit between the simulated and measured values under the conditions of the evaluation, the simulated values cannot be expected to match the measured values outside the current evaluation.

A sample correlation coefficient with a value of 0 indicates no association, and a value of 1 indicates a positive association between the measured and simulated values, but at what value of the sample correlation coefficient do the measured and simulated values stop having no association, and start to be significantly associated?

Many workers use as a standard the squared value of the sample correlation coefficient of 0.8 as an indicator of significant association, that is,

$$r^2 = 0.8.$$

However, the significance of the value of r changes with the number of measured and simulated values compared. A more rigorous approach uses the F test to determine the significance of r (Smith *et al.*, 1996; Smith *et al.*, 1997). The value of F can be calculated using the following formula.

> **Significance of association** between measured and simulated values
> $$F = \frac{(n-2) \times r^2}{1 - r^2},$$
> where n is the number of measured and simulated pairs being compared, and r is the sample correlation coefficient.

The value of F can be related to the probability that the measured and simulated values are not associated by a comparison with the P values given in tables of the F distribution. The P value is multiplied by 100 to give the probability that the measured and simulated values are not associated. Alternatively, the square root of the value of F can be compared to tables of the distribution of the Student's t.

This section has given an overview of methods that can be widely used to quantitatively evaluate the performance of a model with respect to measured values. An example

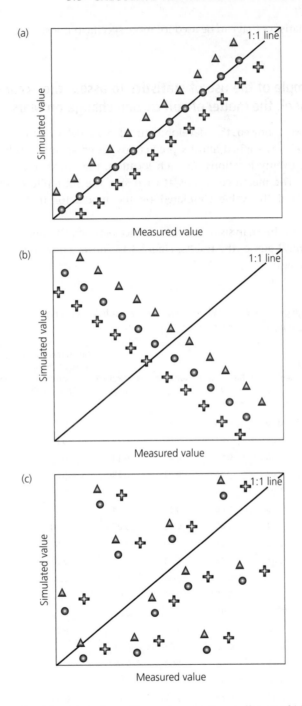

Figure 3.14 Measured and simulated values with sample correlation coefficients of (a) $r = 1$, (b) $r = -1$ and (c) $r = 0$. Circles represent dataset 1; triangles represent dataset 2; crosses represent dataset 3.

of some other statistics that can be used for more specific purposes is given at **Web link 3.2**, box 3.a.

3.3.3 An example of the use of statistics to assess the accuracy of a model: the model of soil carbon change on Mars

In Chapter 2 we followed the development of a model of soil carbon change on Mars. The model was applied and its predictions tested against results collected from robot probes at twenty locations. At each location, the probes took six samples from each treatment. The measured means at each site, the six replicate values that make up the means and the value simulated by the model for that site are shown in Table 3.1.

The first stage in the analysis of the coincidence between the simulated and measured values is a graphical plot of the results. Figure 3.15 shows a plot of the simulated versus the measured values.

Table 3.1 The measured change per year in soil carbon at twenty sites on fields growing Martian fern after twenty years of conversion to zero tillage production, and modelled values.

Site number	Replicate 1 (O_{ij})	Replicate 2 (O_{ij})	Replicate 3 (O_{ij})	Replicate 4 (O_{ij})	Replicate 5 (O_{ij})	Replicate 6 (O_{ij})	Mean measured carbon change value $tCha^{-1}$ (Oi)	Standard error about measured mean	Number of replicates	Modelled value $tCha^{-1}$ (P_i)
1	1.01	0.43	0.90	0.34	0.74	1.03	0.74	0.12	6	0.75
2	0.51	0.17	0.61	0.61	1.04	0.45	0.57	0.12	6	0.64
3	1.04	0.62	0.46	0.87	0.80	0.17	0.66	0.13	6	0.66
4	0.80	0.31	0.68	0.16	0.51	0.73	0.53	0.10	6	0.60
5	0.50	0.46	1.07	0.55	0.34	0.93	0.64	0.12	6	0.70
6	0.88	0.64	0.85	0.46	0.15	0.24	0.54	0.12	6	0.49
7	0.96	0.60	0.34	0.46	0.38	0.27	0.50	0.10	6	0.62
8	0.62	0.31	0.15	0.51	1.00	0.99	0.60	0.14	6	0.69
9	1.06	0.82	0.67	1.02	1.04	0.37	0.83	0.11	6	0.64
10	1.05	0.27	0.67	0.41	0.68	0.44	0.59	0.11	6	0.53
11	1.06	1.01	0.19	0.12	0.58	1.03	0.67	0.18	6	0.68
12	0.70	0.60	0.91	0.95	0.26	0.76	0.70	0.10	6	0.71
13	0.96	0.77	1.02	0.77	0.87	0.99	0.89	0.05	6	0.75
14	0.11	0.66	0.42	0.92	0.67	0.11	0.48	0.13	6	0.49
15	0.62	0.21	0.55	0.18	1.08	0.69	0.55	0.14	6	0.48
16	0.62	0.65	0.98	0.60	0.47	0.56	0.65	0.07	6	0.64
17	0.40	0.92	0.97	0.71	0.71	0.14	0.64	0.13	6	0.51
18	1.03	0.40	0.78	0.40	0.52	0.92	0.67	0.11	6	0.70
19	1.08	0.62	0.79	0.58	0.14	0.13	0.56	0.15	6	0.58
20	1.06	0.94	0.90	0.35	0.63	0.53	0.74	0.11	6	0.69

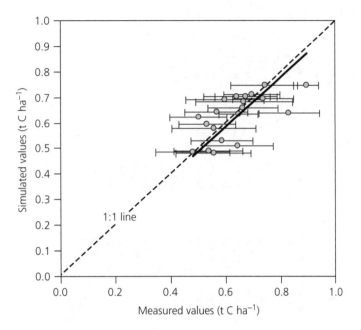

Figure 3.15 The simulated versus the measured change per year in soil carbon at twenty sites on fields growing Martian fern after twenty years of conversion to zero tillage production, and modelled values. The standard errors are shown by the error bars.

The trend between the simulated and measured values is shown by the thick solid line. The 1:1 line (1 unit increase in measured value with respect to 1 unit increase in simulated value) is shown by the dashed line. The proximity of the trend and the 1:1 lines, as well as the overlap between the error bars and the 1:1 line, suggest that the simulations are highly coincident and associated with the measurements; but is this statistically significant?

The degree of coincidence between the simulations and the measurements can be calculated using the methods described in Section 3.3.1 and summarised in the scheme given in Fig. 3.13. The degree of association can be calculated using the methods described in Section 3.3.2.

Replicate values are available, so the best statistic to describe the degree of coincidence is the lack of fit statistic, LOFIT. Table 3.2 shows the calculation of LOFIT, the F value associated with this LOFIT, and the value of F at 5% probability (the critical value below which the degree of coincidence can be considered to be statistically significant). The value of LOFIT is 0.77, which corresponds to a value of F of 0.022. This is less than the critical F value at $P = 0.05$ (critical $F = 1.66$). A value smaller than the critical F value indicates that the lack of fit is not significant, that is, the model error is not greater than the error in the measurements.

If only standard errors and no replicate values had been available (as is often the case when evaluating a model against measurements published in the literature), then the coincidence between the simulated values could be calculated as the root mean squared

Table 3.2 The calculation of LOFIT and F values for simulations of the change per year in soil carbon at twenty sites on fields growing Martian fern after twenty years of conversion to zero tillage production, and modelled values.

Repli-cate 1 (O_{ij})	Repli-cate 2 (O_{ij})	Repli-cate 3 (O_{ij})	Repli-cate 4 (O_{ij})	Repli-cate 5 (O_{ij})	Repli-cate 6 (O_{ij})	Mean measured carbon change value (t C ha^{-1}) (O_i)	Number of replicates (m_i)	Modelled value t C ha^{-1} (P_i)	$O_i - P_i$	$(O_i - P_i)^2$	$m_i(O_i - P_i)^2$	$(m_i - 1)m_i(O_i - P_i)^2$	$\sum_{j=1}^{m_i}((O_{ij} - P_i) - (O_i - P_i))^2$
1.01	0.43	0.90	0.34	0.74	1.03	0.74	6	0.75	−0.00555	0.00003	0.00018	0.00092	0.44083
0.51	0.17	0.61	0.61	1.04	0.45	0.57	6	0.64	−0.07690	0.00591	0.03548	0.17740	0.39869
1.04	0.62	0.46	0.87	0.80	0.17	0.66	6	0.66	0.00411	0.00002	0.00010	0.00051	0.49647
0.80	0.31	0.68	0.16	0.51	0.73	0.53	6	0.60	−0.06376	0.00407	0.02439	0.12195	0.32435
0.50	0.46	1.07	0.55	0.34	0.93	0.64	6	0.70	−0.06468	0.00418	0.02510	0.12551	0.41982
0.88	0.64	0.85	0.46	0.15	0.24	0.54	6	0.49	0.04581	0.00210	0.01259	0.06295	0.46829
0.96	0.60	0.34	0.46	0.38	0.27	0.50	6	0.62	−0.12419	0.01542	0.09254	0.46271	0.31608
0.62	0.31	0.15	0.51	1.00	0.99	0.60	6	0.69	−0.09428	0.00889	0.05333	0.26664	0.60392
1.06	0.82	0.67	1.02	1.04	0.37	0.83	6	0.64	0.19146	0.03666	0.21995	1.09974	0.37087
1.05	0.27	0.67	0.41	0.68	0.44	0.59	6	0.53	0.05454	0.00297	0.01785	0.08923	0.38141
1.06	1.01	0.19	0.12	0.58	1.03	0.67	6	0.68	−0.01601	0.00026	0.00154	0.00769	0.93325
0.70	0.60	0.91	0.95	0.26	0.76	0.70	6	0.71	−0.01674	0.00028	0.00168	0.00841	0.30875
0.96	0.77	1.02	0.77	0.87	0.99	0.89	6	0.75	0.14614	0.02136	0.12813	0.64067	0.06119
0.11	0.66	0.42	0.92	0.67	0.11	0.48	6	0.49	−0.00528	0.00003	0.00017	0.00084	0.54585
0.62	0.21	0.55	0.18	1.08	0.69	0.55	6	0.48	0.07460	0.00556	0.03339	0.16694	0.54823
0.62	0.65	0.98	0.60	0.47	0.56	0.65	6	0.64	0.00967	0.00009	0.00056	0.00280	0.14961
0.40	0.92	0.97	0.71	0.71	0.14	0.64	6	0.51	0.13245	0.01754	0.10526	0.52628	0.50774
1.03	0.40	0.78	0.40	0.52	0.92	0.67	6	0.70	−0.02981	0.00089	0.00533	0.02666	0.38232
1.08	0.62	0.79	0.58	0.14	0.13	0.56	6	0.58	−0.02263	0.00051	0.00307	0.01536	0.69110
1.06	0.94	0.90	0.35	0.63	0.53	0.74	6	0.69	0.04150	0.00172	0.01033	0.05166	0.37439

$$\text{LOFIT} = \sum_{i=1}^{n} m_i(O_i - P_i)^2 = 0.77097$$

$$\sum_{i=1}^{n}(m_i - 1) \times \text{LOFIT} = 3.85486$$

$$\sum_{i=1}^{n}\sum_{j=1}^{m_i}((O_{ij} - P_i) - (O_i - P_i))^2 = 8.72314$$

$$\frac{\dfrac{\sum_{i=1}^{n}(m_i - 1) \times \text{LOFIT}}{n\sum_{i=1}^{n}\sum_{j=1}^{m_i}((O_{ij} - P_i) - (O_i - P_i))^2}} = 0.02210$$

$$F \text{ value } (P = 0.05) = 1.65868$$

Table 3.3 The calculation of RMSE and RMSE$_{95}$ for simulations of the change per year in soil carbon at twenty sites on fields growing Martian fern after twenty years of conversion to zero tillage production, and modelled values.

Modelled value $tCha^{-1}$ (P_i)	Mean measured carbon change value $tCha^{-1}$ (O_i)	Standard error about measured mean	Number of replicates	$(O_i - P_i)^2$	$t_{m,95}$	$(SE_i \times t_{m,95})^2$
0.75	0.74	0.12	6	0.00003	2.77645	0.11327
0.64	0.57	0.12	6	0.00591	2.77645	0.10244
0.66	0.66	0.13	6	0.00002	2.77645	0.12757
0.60	0.53	0.10	6	0.00407	2.77645	0.08334
0.70	0.64	0.12	6	0.00418	2.77645	0.10787
0.49	0.54	0.12	6	0.00210	2.77645	0.12033
0.62	0.50	0.10	6	0.01542	2.77645	0.08122
0.69	0.60	0.14	6	0.00889	2.77645	0.15518
0.64	0.83	0.11	6	0.03666	2.77645	0.09530
0.53	0.59	0.11	6	0.00297	2.77645	0.09800
0.68	0.67	0.18	6	0.00026	2.77645	0.23980
0.71	0.70	0.10	6	0.00028	2.77645	0.07933
0.75	0.89	0.05	6	0.02136	2.77645	0.01572
0.49	0.48	0.13	6	0.00003	2.77645	0.14026
0.48	0.55	0.14	6	0.00556	2.77645	0.14087
0.64	0.65	0.07	6	0.00009	2.77645	0.03844
0.51	0.64	0.13	6	0.01754	2.77645	0.13047
0.70	0.67	0.11	6	0.00089	2.77645	0.09824
0.58	0.56	0.15	6	0.00051	2.77645	0.17758
0.69	0.74	0.11	6	0.00172	2.77645	0.09620

$$\overline{O} = 0.64 \qquad \sum_{i=1}^{n}(O_i - P_i)^2 = 0.12850$$

$$RMSE = \frac{100}{\overline{O}} \times \sqrt{\frac{\sum_{i=1}^{n}(O_i - P_i)^2}{n}} = 12.58673$$

$$\sum_{i=1}^{n}(SE_i \times t_{m,95})^2 = 2.24146$$

$$RMSE_{95} = \frac{100}{\overline{O}} \sqrt{\frac{\sum_{i=1}^{n}(SE_i \times t_{m,95})^2}{n}} = 52.56959$$

error (RMSE). The calculation of RMSE and the value of RMSE at the 95% confidence interval (RMSE$_{95}$) are shown in Table 3.3.

The value of RMSE is 12.59%, which is less than RMSE$_{95}$ (52.57%). This indicates that the total error in the simulation is less than the total error in the measurements at the 95% confidence interval, so confirming the result of the LOFIT calculation.

In the absence of replicates, further information can be given about the nature of the errors by calculating the bias in the error as the relative error, E. The calculation of the relative error, E, and the value of E at the 95% confidence interval (E_{95}) are given in Table 3.4.

Table 3.4 The calculation of E and E_{95} for simulations of the change per year in soil carbon at twenty sites on fields growing Martian fern after twenty years of conversion to zero tillage production, and modelled values.

Modelled value (t C ha^{-1}) (P_i)	Mean measured carbon change value (t C ha^{-1}) (O_i)	$O_i - P_i$	Standard error about measured mean	Number of replicates	$t_{m,95}$	$SE_i \times t_{m,95}$
0.75	0.74	−0.01	0.12	6	2.78	0.34
0.64	0.57	−0.08	0.12	6	2.78	0.32
0.66	0.66	0.00	0.13	6	2.78	0.36
0.60	0.53	−0.06	0.10	6	2.78	0.29
0.70	0.64	−0.06	0.12	6	2.78	0.33
0.49	0.54	0.05	0.12	6	2.78	0.35
0.62	0.50	−0.12	0.10	6	2.78	0.28
0.69	0.60	−0.09	0.14	6	2.78	0.39
0.64	0.83	0.19	0.11	6	2.78	0.31
0.53	0.59	0.05	0.11	6	2.78	0.31
0.68	0.67	−0.02	0.18	6	2.78	0.49
0.71	0.70	−0.02	0.10	6	2.78	0.28
0.75	0.89	0.15	0.05	6	2.78	0.13
0.49	0.48	−0.01	0.13	6	2.78	0.37
0.48	0.55	0.07	0.14	6	2.78	0.38
0.64	0.65	0.01	0.07	6	2.78	0.20
0.51	0.64	0.13	0.13	6	2.78	0.36
0.70	0.67	−0.03	0.11	6	2.78	0.31
0.58	0.56	−0.02	0.15	6	2.78	0.42
0.69	0.74	0.04	0.11	6	2.78	0.31

$$\overline{O} = 0.64 \quad \sum_{i=1}^{m}(O_i - P_i) = 0.18 \qquad\qquad \sum_{i=1}^{n}(SE_i \times t_{m,95}) = 6.53$$

$$E = \frac{100}{\overline{O}} \times \frac{\sum_{i=1}^{m}(O_i - P_i) = 0.18}{n} = 1.42 \qquad E_{95} = \frac{100}{\overline{O}} \frac{\sum_{i=1}^{m}(SE_i \times t_{m,95})}{n} = 51.25$$

The value of E is 1.42%, which is less than E_{95} (51.25%). This indicates that the bias in the error is not significant.

If neither replicates nor standard errors are given (as can be the case when using data from a long-term experiment or an expensive field trial), then the coincidence between the simulated values can still be evaluated against an acceptable error. An RMSE value of less than 10% is often considered to be acceptable, so the calculated RMSE of 12.59% is well within the acceptable error. The significance of the bias can be defined by calculating the mean difference, M, and the associated t value, as shown in Table 3.5. The value of M is 0.01, which corresponds to a t value of 0.51. This is less than the critical value of the Student's t at a P value of 5%, namely, 2.09, suggesting that there is no significant bias in the simulation.

Table 3.5 The calculation of M and t for simulations of the change per year in soil carbon at twenty sites on fields growing Martian fern after twenty years of conversion to zero tillage production, and modelled values.

Modelled value (t C ha^{-1} (P_i))	Mean measured carbon change value (t C ha^{-1}) (O_i)	($O_i - P_i$)	$\left(O_i - P_i - \left(\sum_{i=1}^{n}(O_i - P_i)/n\right)\right)^2$
0.75	0.74	−0.01	0.00
0.64	0.57	−0.08	0.01
0.66	0.66	0.00	0.00
0.60	0.53	−0.06	0.01
0.71	0.64	−0.07	0.01
0.49	0.54	0.05	0.00
0.62	0.50	−0.12	0.02
0.69	0.60	−0.09	0.01
0.64	0.83	0.19	0.03
0.53	0.59	0.06	0.00
0.68	0.67	−0.02	0.00
0.71	0.70	−0.02	0.00
0.75	0.89	0.15	0.02
0.49	0.48	−0.01	0.00
0.48	0.56	0.08	0.00
0.64	0.65	0.01	0.00
0.51	0.64	0.13	0.02
0.70	0.68	−0.03	0.00
0.58	0.56	−0.02	0.00
0.69	0.74	0.04	0.00

$$M = \frac{\sum_{i=1}^{n}(O_i - P_i)}{n} = 0.01$$

$$\sum_{i=1}^{n}\left(O_i - P_i - \left(\sum_{i=1}^{n}(O_i - P_i)/n\right)\right)^2 = 0.13$$

$$t \text{ value} = \frac{M \times \sqrt{n}}{\sqrt{\frac{\sum_{i=1}^{n}\left(O_i - P_i - \left(\sum_{i=1}^{n}(O_i - P_i)/n\right)\right)^2}{(n-1)}}} = 0.51$$

The association between the simulated and the measured values is calculated using the correlation coefficient, r, and its significance is determined using an F test. The calculations of r and F are shown in Table 3.6.

The correlation coefficient, r, has a value of 0.66. The F value associated with this value of r is 14.15. The F value is greater than the critical F value at $P = 0.05$, indicating that the association between the modelled and the measured values is significant and suggesting a good match between the patterns in the simulations and the measurements.

Table 3.6 The calculation of r, F and the critical F at $P = 0.05$ for simulations of the change per year in soil carbon at twenty sites on fields growing Martian fern after twenty years of conversion to zero tillage production, and modelled values.

Modelled value (t C ha^{-1}) (P_i)	Mean measured carbon change value (t C ha^{-1}) (O_i)	$P_i - \bar{P}$	$(P_i - \bar{P})^2$	$O_i - \bar{O}$	$(O_i - \bar{O})^2$	$(O_i - \bar{O})(P_i - \bar{P})$
0.75	0.74	0.12	0.01	0.10	0.01	0.012
0.64	0.57	0.02	0.00	-0.07	0.00	-0.001
0.66	0.66	0.03	0.00	0.03	0.00	0.001
0.60	0.53	-0.03	0.00	-0.10	0.01	0.003
0.70	0.64	0.08	0.01	0.00	0.00	0.000
0.49	0.54	-0.14	0.02	-0.10	0.01	0.014
0.62	0.50	0.00	0.00	-0.14	0.02	0.000
0.69	0.60	0.06	0.00	-0.04	0.00	-0.003
0.64	0.83	0.01	0.00	0.19	0.04	0.002
0.53	0.59	-0.10	0.01	-0.05	0.00	0.005
0.68	0.67	0.05	0.00	0.03	0.00	0.002
0.71	0.70	0.08	0.01	0.06	0.00	0.005
0.75	0.89	0.12	0.01	0.26	0.07	0.031
0.49	0.48	-0.14	0.02	-0.16	0.02	0.022
0.48	0.55	-0.15	0.02	-0.08	0.01	0.012
0.64	0.65	0.01	0.00	0.01	0.00	0.000
0.51	0.64	-0.12	0.01	0.01	0.00	-0.001
0.70	0.67	0.08	0.01	0.04	0.00	0.003
0.58	0.56	-0.05	0.00	-0.08	0.01	0.004
0.69	0.74	0.07	0.00	0.10	0.01	0.007

$\bar{P} = 0.63$ $\bar{O} = 0.64$ $\sum_{i=1}^{n} (P_i - \bar{P})^2 = 0.15$ $\sum_{i=1}^{n} (O_i - \bar{O})^2 = 0.22$ $\sum_{i=1}^{n} (O_i - \bar{O})$

$$\sqrt{\sum_{i=1}^{n} (P_i - \bar{P})^2} = 0.15 \qquad \sqrt{\sum_{i=1}^{n} (O_i - \bar{O})^2} = 0.47 \qquad (P_i - \bar{P}) = 0.12$$

$$r = \frac{\sum_{i=1}^{n} (O_i - \bar{O})(P_i - \bar{P})}{\sqrt{\sum_{i=1}^{n} (O_i - \bar{O})^2} \sqrt{\sum_{i=1}^{n} (P_i - \bar{P})^2}} = 0.66$$

$$F = \frac{(n-2) \times r^2}{1 - r^2} = 14.15$$

$$F \text{ value } (P = 0.05) = 4.41$$

The above statistical analysis confirms the observations from the graphical plot of simulated versus the measured values (Fig. 3.15). The results indicate satisfactory model performance with respect to both coincidence and association, suggesting that the model can reliably be used to simulate the changes per year in soil carbon following conversion to zero tillage on the range of conditions covered by sites 1 to 20. We can now use the model in these conditions with confidence.

SELF-CHECK QUESTIONS: HOW WOULD YOU QUANTIFY THE ACCURACY OF THE MODEL?

1. Your model of changes in soil carbon on Mars is tested against some independent data recently collected during fresh missions by robot probes that landed on Mars.

   Q: Which statistics would you use to check the degree of *association* between the measured and simulated values?
   a. Relative error
   b. Mean difference
   c. Correlation coefficient
   d. Root mean squared error
   e. Lack of fit (LOFIT)

   [A: c.]

   Q: Which statistics would you use to check the bias in your model?
   a. Relative error
   b. Mean difference
   c. Correlation coefficient
   d. Root mean squared error
   e. Lack of fit (LOFIT)

   [A: a or b.]

   Q: If you have replicate measurements, which statistics would you use to look at error in your model?
   a. Relative error
   b. Mean difference
   c. Correlation coefficient
   d. Root mean squared error
   e. Lack of fit (LOFIT)

   [A: e.]

   Q: If you *do not* have replicate measurements, which statistics would you use to look at error in your model?
   a. Relative error
   b. Mean difference
   c. Correlation coefficient
   d. Root mean squared error
   e. Lack of fit (LOFIT)

   [A: d.]

3.4 **Examine the behaviour of the model (sensitivity analysis)**

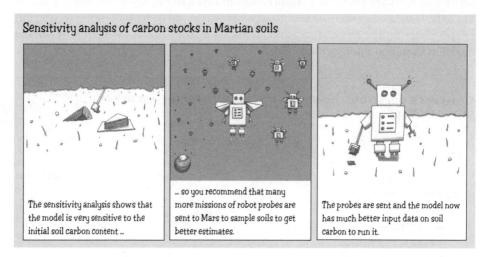

Sensitivity analysis of carbon stocks in Martian soils

The sensitivity analysis shows that the model is very sensitive to the initial soil carbon content ...

... so you recommend that many more missions of robot probes are sent to Mars to sample soils to get better estimates.

The probes are sent and the model now has much better input data on soil carbon to run it.

3.4.1 **What is sensitivity analysis and why is it important?**

Sensitivity analysis identifies which model components exert the most influence on the model results. It compares changes in the simulated values against changes in the model components; these components could be input variables or fixed parameters. If the validity of the model response is assessed in a qualitative way by expert judgement, then this is usually referred to as a **subjective sensitivity analysis** (Hamby, 1994) or a **logical analysis**. If the assessment of the model response is quantitative, this constitutes a **quantitative sensitivity analysis**.

Sensitivity analysis is closely related to **uncertainty analysis** (see Section 3.5). Sensitivity analysis determines how highly correlated the model result is to the value of a given input component—does a small change in the input cause a significant change in the output? If this is the case, the model is termed **sensitive** to the input. An uncertainty analysis determines how the **variability** in the input is propagated through the model and quantifies how this is translated into variability (uncertainty) in the model output. If this is the case, the input is termed **important**. A sensitivity analysis and an uncertainty analysis do not necessarily identify the same inputs. A model is always **sensitive** to the **important** inputs that contribute most to model output uncertainty, since the variability will not appear in the model output unless the model is also sensitive to the input. However, an input to which the model is **sensitive** is not necessarily **important**; it does not necessarily contribute to output uncertainty since its value may be known precisely. Hamby (1994) and Saltelli *et al.* (2000) provide excellent further reading on sensitivity analysis.

There are many techniques that can be used to perform a sensitivity analysis and all rely upon examining changes in model outputs when input components are varied. Some techniques are described in the following section.

3.4.2 Methods used in sensitivity analysis

In its simplest form, a sensitivity analysis entails adjustment of the model input components one at a time, whilst all others remain constant, and the influence of each input on the model outputs is examined. This form of sensitivity analysis is termed, not surprisingly, **one-at-a-time** or **local** sensitivity analysis (see **Web link 3.2**, box 3.b). The input can be varied by an arbitrary amount (e.g., by 50% of the estimated or mean value), or by a function of its variability, if known (e.g., by a function of its standard deviation). If the variability of the input is known or can be estimated, then the actual range of the input can be entered.

A sensitivity analysis can also involve adjustment of more than one input at a time. In its simplest form, a multiple sensitivity analysis takes the form of a **factorial analysis**, which involves choosing a given number of samples for each input and running the model for all combinations. Computationally, both a one-at-a-time and a factorial analysis can be performed by a **grid search**, whereby a grid of values is used to run the model for multiple runs.

This type of sensitivity analysis, in which the effect of varying all inputs simultaneously is examined, is known as a **global sensitivity analysis**. A global sensitivity analysis has the advantage of assessing input sensitivity in the context of other varying inputs. If the ranges and distributions of the inputs are known or can be measured, a **probability density function** can be defined for each input. The probability density function gives the distribution of the possible values of an input, and so can be used to randomly select input values from within these distributions (see Section 3.5), thereby allowing the full range of potential model outputs to be examined. This can be used to define input sensitivity as described for one-at-a-time sensitivity analysis, by comparing how changes in the inputs affect the model results. If the variation in the outputs is compared to the variation in the inputs, a global uncertainty analysis is performed (see Section 3.5). There are many software packages, such as SimLab (**Web link 3.3**) and MATLAB (**Web link 3.4**), that allow numerous variations on a one-at-a-time and a global sensitivity or uncertainty analysis to be performed.

3.4.3 Expressing sensitivity

Sensitivity can be determined qualitatively by plotting the outputs against the inputs. A very convenient quantitative expression of sensitivity is then given by the correlation coefficient between the inputs and the outputs.

> **Quantitative expression of model sensitivity** to variation in one input (assuming a linear relationship between the inputs and the outputs)
>
> For example, sample correlation coefficient, $r = \dfrac{\sum\limits_{i=1}^{n} (I_i - \bar{I})\,(P_i - \bar{P})}{\sqrt{\sum\limits_{i=1}^{n} (I_i - \bar{I})^2}\;\sqrt{\sum\limits_{i=1}^{n} (P_i - \bar{P})^2}},$
>
> where I_i is the ith input value, $\bar{I}$ is the average measured value, P_i is the ith output value, and $\bar{P}$ is the average output value.

An expression of sensitivity to a variation in more than one input at a time can be calculated using a regression analysis between the inputs and the outputs.

A major drawback of using the correlation coefficient to define sensitivity is the inherent assumption that the relationship between the inputs and the outputs is linear. Also, the possibility that the input parameters are strongly correlated to each other can result in apparent correlations between the inputs and the outputs that do not exist. If a nonlinear relationship between the inputs and the outputs or a correlation with other inputs is likely, then other less powerful methods of expressing sensitivity must be used.

Other ways of expressing the sensitivity include the **sensitivity coefficient**, which is the ratio of the change in the output to the change in one input whilst all other inputs remain constant. The model outputs are compared to the base case output, which is the model result with all inputs held constant (Hamby, 1994). This gives the sensitivity of the model, but only to a fixed change in a single input.

> **Quantitative expression of model sensitivity** to a fixed change in a given input
>
> $$\text{Sensitivity coefficient} = \frac{P_i - \bar{P}}{I_i - \bar{I}},$$
>
> where I_i is the ith input value, $\bar{I}$ is the average measured value, P_i is the ith output value, and $\bar{P}$ is the average output value.

The sensitivity of the model over the entire possible range of an input can be expressed using the **sensitivity index**. By varying the parameter from its minimum to its maximum and examining the minimum and maximum output values, the sensitivity index for a parameter is given by the maximum output value minus the minimum output value, all divided by the maximum output value.

> **Quantitative expression of model sensitivity** to a range of changes in a given input
>
> $$\text{Sensitivity index} = \frac{\max(P_i) - \min(P_i)}{\max(P_i)},$$
>
> where $\max(P_i)$ is the maximum output value, and $\min(P_i)$ is the minimum output value from the range of the input value used.

A range of other approaches to expressing sensitivity exists. Hamby (1994) and Saltelli *et al.* (2000) have reviewed many of these approaches.

SELF-CHECK QUESTIONS: HOW WOULD YOU QUANTIFY THE BEHAVIOUR OF THE MODEL?

1. Q: What does a sensitivity analysis tell you about a model input?
   a. How a change in the value of the model input causes a change in the value of the output
   b. How the *variability* in the input is propagated through the model and quantifies how this is translated into variability in the model output

   [A: a (b is an uncertainty analysis).]

2. Q: What is the difference between a local and a global sensitivity analysis?
   a. A local sensitivity analysis is carried out on site-specific models and a global sensitivity analysis is carried out on global models
   b. A local sensitivity analysis varies one input at a time whilst a global sensitivity analysis holds one parameter constant but varies all other parameters simultaneously

   [A: b.]

3. Q: Which of the following measures are often used to express model sensitivity?
   a. Relative error
   b. Mean difference
   c. Sample correlation coefficient
   d. Root mean squared error
   e. Lack of fit (LOFIT)
   f. Sensitivity coefficient
   g. Sensitivity index

   [A: c, f and g.]

3.5 Determine the importance of the model components (uncertainty analysis)

3.5.1 What is uncertainty analysis and why is it important?

An uncertainty analysis identifies the model components for which variability in the inputs exerts the greatest influence on variability, or uncertainty, of the model outputs.

Uncertainty analysis is related to **sensitivity analysis** (see Section 3.4). A sensitivity analysis determines if model results are correlated with a given input, that is, if a small change in the input results in a significant change in the output. In this case, a model is termed **sensitive** to the input. An **uncertainty analysis** determines if the **variability** in the input is propagated through the model and leads to a high variability (i.e., uncertainty) in the model outputs. In this case, the input is termed **important**. A sensitivity analysis and an uncertainty analysis do not necessarily identify the same input components. **Important** inputs (i.e., those contributing most to model output uncertainty) are always the inputs that the model is **sensitive** to, since the variability will not appear in the model output unless the model is sensitive to that input. However, inputs to which the model is **sensitive** may be known precisely, so do not necessarily contribute to output uncertainty. Hamby (1994) provides excellent further reading on uncertainty analysis.

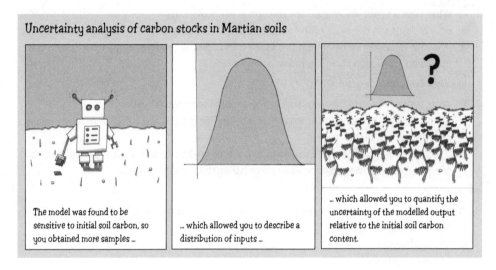

Uncertainty analysis of carbon stocks in Martian soils

The model was found to be sensitive to initial soil carbon, so you obtained more samples ...

... which allowed you to describe a distribution of inputs ...

... which allowed you to quantify the uncertainty of the modelled output relative to the initial soil carbon content.

3.5.2 Methods used in uncertainty analysis

The simplest form of uncertainty analysis, like sensitivity analysis, is to investigate the influence of the variation in the model inputs one at a time, whilst all other inputs remain constant, and examine the influence of the variability of each input on the model output variability. This is analogous to a **one-at-a-time**, or a **local**, sensitivity analysis. If the variability of the input is known or can be estimated, then an uncertainty analysis can be conducted. For this, the input parameter is varied within the range of its statistical distribution (defined, for example, by its standard deviation or by its 95% confidence interval about the mean value) and the variability in the model output is measured.

More complex forms of uncertainty analysis involve using partial differentiation of the model in its aggregated form, often called **differential analysis** (Hamby, 1994). The mathematics required to perform a differential analysis is beyond the scope of this book, but if you want to know more, this is dealt with in some detail by Hamby (1994) and Saltelli *et al.* (2000).

3.5.3 Representing variation in the input parameters and model outputs

The variation in the input factor can be represented in many ways, for example by using fixed percentiles about the mean, by multiplications of the standard deviation, by randomly sampling different distributions, or by defining **probability density functions** (the potential distribution). If the mean, variance and distribution shape for the inputs are known, then the probability density function can be defined. The probability density function gives the distribution of the input with respect to its mean value. Random sampling of the input distribution can then be used to examine the importance of inputs on the uncertainty in the model outputs. **Monte Carlo** techniques randomly sample from within the potential distribution of the input values. For greater computational efficiency, **Latin hypercube** and **Plackett–Burnham** techniques are mathematical ways of simplifying this process to approximate the Monte Carlo technique whilst drastically reducing the number of model runs required. There are many software packages, such as SimLab (**Web link 3.3**) and MATLAB (**Web link 3.4**), that allow numerous variations on a one-at-a-time and a global sensitivity or uncertainty analysis to be performed. Variation in the model output can be represented in similar ways.

3.5.4 Expressing uncertainty

As for sensitivity, uncertainty can be determined qualitatively by plotting the variation in the inputs against the variation in the outputs, or can be expressed quantitatively by calculating the correlation coefficient or by a regression analysis.

Using the correlation coefficient to define uncertainty requires that the relationship between the inputs and the outputs is approximately linear. A less powerful statistic, but one that does not assume a linear relationship, is the **importance index**. This gives a measure of the fractional contribution of the input to the overall output variability or uncertainty. It is given by the variance of the input value divided by the variance of the model output, and can be resolved into the following equation.

Quantitative expression of importance of one input value in determining the uncertainty of the model output

$$\text{Importance index} = \frac{\sum_{i=1}^{n}(I_i - \bar{I})^2}{\sum_{i=1}^{n}(P_i - \bar{P})^2},$$

where I_i is the ith input value, $\bar{I}$ is the average measured value, P_i is the ith output value, and $\bar{P}$ is the average output value.

The **relative deviation method** ranks uncertainty by the amount of variability introduced into the model output whilst varying each input one at a time according to its probability density function. The relative deviation is the ratio of the standard deviation to the mean of the output density function, that is,

$$\text{Relative deviation ratio} = \frac{\sqrt{\sum_{i=1}^{n}(P_i - \bar{P})^2/(n-1)}}{\bar{P}},$$

where P_i is the ith output value, $\overline{P}$ is the average output value, and n is the total number of values in the sample.

The ratio of the relative deviation of the output to the relative deviation of the inputs can then be used to rank the sensitivity of the model to different inputs. This can be resolved into the following equation.

Quantitative expression of importance of one input value in determining the uncertainty of the model output

$$\text{Relative deviation ratio} == \frac{\overline{O} \times \sqrt{\sum_{i=1}^{n} (P_i - \overline{P})^2}}{\overline{P} \times \sqrt{\sum_{i=1}^{n} (O_i - \overline{O})^2}},$$

where P_i is the ith output value, $\overline{P}$ is the average output value, n is the total number of values in the sample, O_i is the ith input value, and $\overline{O}$ is the average input value.

Other methods of expressing uncertainty include **rank transformations**, which allow for nonlinearities in the data, the **partial correlation coefficient**, which allows correlation among input variables to be accounted for, and **regression analyses** which can be used to plot response surfaces of the model outputs to the most sensitive model parameters. Hamby (1994) describes all of these measures and techniques (and others) in detail.

SELF-CHECK QUESTIONS: HOW WOULD YOU QUANTIFY THE IMPORTANCE OF THE MODEL COMPONENTS?

1. Q: What does an uncertainty analysis tell you about a model input?
   a. How a change in the value of the model input causes a change in the value of the output
   b. How the *variability* in the input is propagated through the model and quantifies how this is translated into variability in the model output

   [A: b (a is a sensitivity analysis).]

2. Q: What is a probability density function?
   a. The probability that a given input value is correct
   b. A distribution of potential values for a model input
   c. A measure of model uncertainty

   [A: b.]

3. Q: Which of the following measures are often used to express model uncertainty?
   a. Sample correlation coefficient
   b. Importance index
   c. Relative deviation ratio
   d. Root mean squared error
   e. Lack of fit (LOFIT)
   f. Sensitivity coefficient
   g. Sensitivity index

   [A: b and c.]

3.6 And then . . .?

In this chapter we have described a number of methods that can be used to evaluate the performance of a model. We have discussed how the results should first be plotted to highlight any outliers, a systematic shift in the simulated values with respect to the measured values, or differences in the trends of the simulated and measured values. We have considered other types of plots that allow patterns in errors to be identified, illustrate the behaviour of the model components, and allow the importance of different model components to be established. We have described a procedure for expressing the accuracy of the simulation, by calculating the degree of coincidence and association between the modelled and measured data. We have discussed quantitative methods used to examine the behaviour of the model components and to determine the importance of each component to the uncertainty of the model output. So now, can we get on and use the model?

Yes, at last! The model has been properly developed and evaluated, and so is now ready to use in whatever way you choose! The model can be used unchanged, simply to interpret experimental observations of a system, to improve our understanding of observations, and to identify where observations cannot be explained by the existing model—so furthering our scientific understanding. The model can be constructed into an expert system or a decision support system to help guide non-experts in decision processes that require expert knowledge. The model can be used to assess environmental risks associated with a particular course of action. For larger-scale decisions, the model can be constructed into a geographical information system to provide spatially-explicit information. The model has been evaluated and the confidence and restrictions that should be put on the application of the model have been clearly defined, so the application of the model *should* be straightforward. However, there are further ways that models can be misused and the results abused. The prudent application of models is the subject of the next chapter.

Your model works well — it is time to apply it

Your model gives a good fit against all of the samples taken by the robot probe ...

An uncertainty analysis suggests that the results of your model are likely to be accurate given your input data.

Your model suggests that cultivation of the soil has caused a huge release of CO_2 into the atmosphere, causing climate change.

Your model also suggests that most of the CO_2 added to the atmosphere could be locked up in the soil by changing tillage practice ...

... and adding compost back to the fields.

It is time to let the Greys know what they need to do to solve their problems.

■ **SUMMARY**

Stage 1: Plot the results (graphical analysis)

1. Plots to reveal the accuracy of the simulation should
   a. show the result that is needed from the model,
   b. include the simulated and measured values on the same plot,
   c. show errors in the measurements as error bars, and variations in simulations as error bars or as a band of potential results,
   d. use axes that highlight the acceptable error in the simulation, and
   e. avoid confusing and complicated presentation.

2. Plots to illustrate the behaviour and importance of the model components should
   a. only include as much complexity as can be interpreted by eye,
   b. include any measurements that are available, and
   c. show the importance of the components as changes in the distribution of the results.

Stage 2: Calculate the accuracy of the simulation (quantitative analysis)

1. A thorough quantitative analysis should include both
   a. an analysis of coincidence (difference), and
   b. an analysis of association (trends).

2. A quantitative analysis compares the simulations to the independent measurements.

3. Coincidence can be expressed as
   a. the total difference (root mean squared error),
   b. the bias (relative error or mean difference), or
   c. the error excluding error due to variations in the measurements (lack of fit).

4. The significance of the coincidence can be determined by
   a. a comparison to the 95% confidence interval, or
   b. a direct comparison to the P value obtained in the t test or the F test.

5. The association can be expressed as the sample correlation coefficient.

6. The significance of the association can be determined using a t-test.

Stage 3: Examine the behaviour of the model (sensitivity analysis)

1. Common types of sensitivity analysis are
   a. a one-at-a-time sensitivity analysis (inputs varied one by one), and
   b. a factorial analysis (more than one parameter adjusted at a time).

2. Sensitivity may be expressed (among others)
   a. as the correlation coefficient between the inputs and the outputs,
   b. by regression analysis,
   c. as the sensitivity coefficient, or
   d. as the sensitivity index.

Stage 4: Determine the importance of the model components (uncertainty analysis)

1. Common types of uncertainty analysis are
   a. a one-at-a-time uncertainty analysis, and
   b. a differential analysis.

2. A variation in the input parameters is represented
   a. as a fixed percentile about the mean,
   b. using a probability density function and Monte Carlo sampling.

3. Uncertainty can be expressed (among others)
   a. as the importance index, or
   b. as the relative deviation ratio.

■ PROBLEMS (SOLUTIONS ARE IN APPENDIX 1.3)

3.1 Evaluate the performance of a model that simulates the optimum nitrogen fertiliser application rate to a range of crops. The following table gives the measured optimum

Crop	Yield t ha^{-1}	Simulated value	Measured value					
			Repli-cate 1	Repli-cate 2	Repli-cate 3	Repli-cate 4	Repli-cate 5	Repli-cate 6
Spring wheat	5.1	120	138	130	125	115	112	110
	5.7	120	110	100	114	125	135	117
	6	120	100	105	102	108	110	130
	5.4	100	118	105	95	115	112	85
	5.8	100	115	97	133	122	145	116
	6	100	92	111	91	99	85	122
	5.8	140	145	157	155	132	138	121
	6	120	132	103	131	142	106	121
	5.4	100	87	102	97	112	118	105
	7.1	160	143	165	175	154	152	176
	7.4	150	157	132	165	175	132	151
	7.2	140	132	154	126	121	162	157
	6.5	160	176	165	160	145	147	141
	6.8	140	138	125	129	147	157	165
Winter wheat	6	160	155	150	162	165	167	152
	6.5	159	150	145	157	167	170	175
	7.2	161	160	167	172	189	199	170
	7.4	181	165	200	173	171	168	184
	6.5	141	170	165	175	187	170	175
	7.2	139	170	167	172	152	155	170
	6.7	162	166	155	145	182	187	177
	6.4	141	154	132	161	123	153	125
	7	161	175	153	145	154	180	174
	7.8	181	164	167	178	197	193	178
	8.5	200	186	182	194	214	218	201
	8	180	174	194	204	176	164	177
	7.4	160	162	145	178	153	169	180
	7.2	180	200	202	187	193	197	173
Winter oilseed rape	6	160	155	150	162	165	167	152
	3.5	220	175	165	186	155	180	185
	4	250	200	195	193	205	169	172
	4.2	263	205	210	190	203	202	185
	3.9	250	188	182	177	155	201	200
	4.1	220	181	175	172	203	172	155
	4.2	220	162	125	143	135	133	111
	3.5	220	154	142	135	154	137	175
	3.6	220	176	159	154	171	145	197
	3.2	200	165	127	138	143	176	146
	4	241	201	197	176	186	180	173
	4.7	279	220	243	237	222	254	265
	4.4	260	210	212	254	220	223	201
	2.5	141	121	100	123	75	89	92
	3.2	179	134	156	123	111	142	152

nitrogen fertiliser application rate for a range of different crops. The rate recommended by the model is also given in the table.

Plot the results. What do the plots show? Determine the coincidence and association between the simulated and the measured values. Is the model providing a good estimate of the optimum nitrogen application rate? Discuss how well the model performs overall, and for each individual crop. The MODEVAL spreadsheet (in Microsoft Excel, see **Web link 3.1**) can be used (if required) to calculate the statistics.

3.2 **Examine the behaviour of a simple model** of the growth of a rabbit population in a warren during summer. The number of rabbits is estimated using the following model:

$$R = R_{start} + \exp(k_1 \times t) + \exp(k_2 \times t),$$

where R_{start} is the measured number of rabbits (500 rabbits), t is the time in weeks since the number of rabbits was measured, and k_1 and k_2 are parameters describing the growth in the population ($k_1 = 0.5$ and k_2, 0.02).

Plot the calculated change in the rabbit population with time from zero to ten weeks. Use a grid search to investigate the sensitivity of the model to changes in R_{start}, k_1 and k_2 within the range of $\pm 20\%$ with step sizes of 5%. Plot the results and calculate the sensitivity index for each combination of R_{start}, k_1 and k_2.

Can the model be simplified in any way?

3.3 **Determine the importance of the model components** to the uncertainty of the rabbit population growth model (see Problem 3.2). The number of rabbits can be measured with 20% accuracy, and the population growth parameters are known to be 50% accurate. Define the importance of each input using the importance index and the relative deviation ratio, assuming a uniform distribution for the probability density functions (i.e., equal probability of finding a parameter at each grid point from the mean). What research is needed to reduce the uncertainty of the model?

▨ FURTHER READING

Chatfield, C. (1983). *Statistics for technology* (3rd edn.). Chapman and Hall, London.

Hamby, D. M. (1994). A review of techniques for parameter sensitivity analysis of environmental models. *Environmental Monitoring and Assessment*, **32**, 135–54.

Saltelli, A., Chan, K. and Scott, E. M. (2000). *Sensitivity analysis*. Wiley, Chichester.

Townend, J. (2002). *Practical statistics for environmental and biological scientists*. Wiley, Chichester.

▨ REFERENCES

Addiscott, T. M. and Whitmore, A. P. (1987). Computer simulation of changes in soil mineral nitrogen and crop nitrogen during autumn, winter and spring. *Journal of Agricultural Science, Cambridge*, **109**, 141–57.

Loague, K. and Green, R. E. (1991). Statistical and graphical methods for evaluating solute transport models: overview and application. *Journal of Contamination Hydrology*, **7**, 51–73.

Smith, J. U., Smith, P. and Addiscott, T. M. (1996). Quantitative methods to evaluate and compare soil organic matter (SOM) models. In *Evaluation of soil organic matter models using existing long-term datasets*, NATO ASI Series I, Vol. 38 (ed. D. S. Powlson, P. Smith and J. U. Smith), pp. 81–98.

Smith, P., Smith, J. U., Powlson, D. S., McGill, W. B., Arah, J. R. M., Chertov O. G., *et al.* (1997). A comparison of the performance of nine soil organic matter models using datasets from seven long-term experiments. *Geoderma*, **81**, 153–225.

Smith, P., Powlson, D. S., Smith, J. U., Falloon, P. D. and Coleman, K. (2000). Meeting Europe's climate change commitments: Quantitative estimates of the potential for carbon mitigation by agriculture. *Global Change Biology*, **6**, 525–39.

Whitmore, A. P. (1991). A method for assessing the goodness of computer simulations of soil processes. *Journal of Soil Science*, **42**, 289–99.

■ WEB LINKS

Web link 3.1: Online resource centre: **www.oxfordtextbooks.co.uk/orc/smith_smith/**

Web link 3.2: Online resource centre: **www.oxfordtextbooks.co.uk/orc/smith_smith/**

Web link 3.3: **http://www.simlab.de/**
SimLab home page

Web link 3.4: **http://www.mathworks.com/**
MATLAB home page

Web link 3.5: Online resource centre: **www.oxfordtextbooks.co.uk/orc/smith_smith/**

Web link 3.6: Online resource centre: **www.oxfordtextbooks.co.uk/orc/smith_smith/**

Web link 3.7: Online resource centre: **www.oxfordtextbooks.co.uk/orc/smith_smith/**

4 How to apply a model

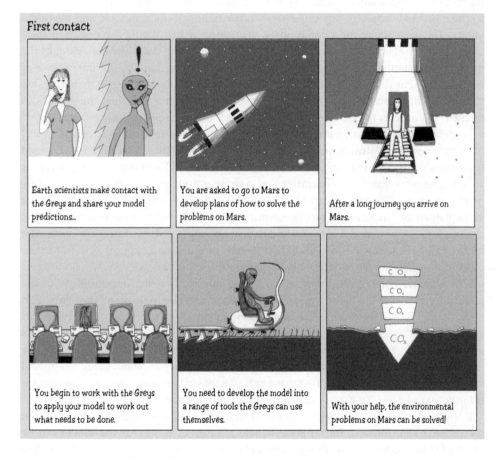

First contact

Earth scientists make contact with the Greys and share your model predictions...

You are asked to go to Mars to develop plans of how to solve the problems on Mars.

After a long journey you arrive on Mars.

You begin to work with the Greys to apply your model to work out what needs to be done.

You need to develop the model into a range of tools the Greys can use themselves.

With your help, the environmental problems on Mars can be solved!

As already discussed in the preceding chapters, making a model that does just what you want it to do involves a series of important stages: model development, evaluation and application.

Model development starts with the research question or practical objective. First, you must decide what type of model to use from the list of the reasons for developing the model. The problem must then be represented as a conceptual model, namely, a picture of the system and the lists of hypotheses, assumptions and boundary conditions. A mathematical model must be formulated around the conceptual model by deriving fixed parameters, input variables and the link between the input variables and the required results. Finally, you must develop the mathematical model into a computer model, using different types of software to set aside computer memory for the state variables, tell the computer how one state variable affects another, and obtain inputs and return outputs to the user in an intelligible format.

Model evaluation can take a number of different forms. It can involve the determination of the accuracy of the simulations, an analysis of the behaviour of the model, or the identification of the important components of the model. The results must first be plotted to highlight any outliers, a systematic shift in the simulated values with respect to the measured values, or any differences in the trends of the simulated and measured values. The accuracy of the simulations is quantified by calculating the degree of coincidence and association between the modelled and measured data. The behaviour of the model components is examined using a sensitivity analysis. Finally, the importance of each component is determined using an uncertainty analysis. If it can be demonstrated that the model is accurate to within the limits of acceptable error, then the model can be used, but only in the environments for which it has been successfully evaluated.

Model application is the last step toward getting some real use out of the model in which you have already invested so much time and effort. A model can be used for a number of different purposes. The model can be used without further packaging to interpret experimental observations of a system and improve our understanding of observations. The model can also be used to identify where observations cannot be explained by the existing model, so furthering our scientific understanding of the system. The use of a model for scientific representation is discussed in Section 4.1. The model can be constructed into an expert system or a decision support system to help guide non-experts in decision processes that require expert knowledge. The non-expert use of a model holds great potential for model misuse. It is the responsibility of the application developer to guard against all potential errors due to incorrect data entry and misinterpretation of the results. The application of a model in expert and decision support systems is discussed in Section 4.2. The model can be used to assess environmental risks associated with a particular course of action. The important role of models in risk assessment is discussed in Section 4.3. For assessing impacts at larger scales, the model can be made spatially explicit, often using at least some of the capabilities of geographical information systems. Spatially-explicit applications are discussed in Section 4.4.

If the model has been properly developed and evaluated, then the model application *should* be straightforward, but as always there are further ways that the model can be misused. The prudent application of a model requires answers to the following five key questions.

1. Who is the end-user?
2. What will the model do?
3. How can the application guard against input error?
4. How can the application guard against misinterpretation of the results?
5. What documentation is needed?

Failure to consider these questions could result in a model that is excellent in its conception and accurate in its construction, but erroneous in its implementation due to the inability of the user to supply the correct inputs, to accurately interpret the results or to get the model to run at all.

The identity of the end-user is of crucial importance in determining what type of user support is needed to run the model; should the model run through a graphical user interface or can the application be driven by simple input files? Developing a sophisticated graphical user interface not only takes time, but can also make the model appear like a 'black box', where even an expert user has difficulty in understanding the processes behind the interface. Graphical user interfaces can obscure the model functions for an expert user, but are essential if the model is to be used by a non-expert. The identity of the end-user determines the most appropriate packaging to put around your model.

How the model will be used further determines the nature of the model packaging; will the model be used to provide a single prescriptive result, or does it provide information that supports complex decisions made by the end-user? How much flexibility is needed in the choice of input data? Ideally, the user should specify inputs that describe the exact conditions of the model run, but if the way the model is being used means a particular input variable always has a similar value, then this input might be fixed within the model to make data entry less cumbersome. How much access to the results is needed? If the model will be used to understand the process controls in the system, detailed results will be required; a more limited form of output will be sufficient for many other processes.

The method used to guard against input error depends on the identity of the end-user. If the end-user is an expert, it might be sufficient to highlight unusual inputs with a warning message, but allow the user to continue. If, however, the end-user is a non-expert, it might be more prudent to disallow unusual inputs using an error message to terminate the model run or to force the user to enter a different value.

The method used to guard against misinterpretation of the results depends on the way the model is used, as well as the identity of the end-user. A detailed graphical and statistical analysis of the results may be appropriate for an expert user who is applying the model to understand the processes occurring in the system, whereas a single number output is more appropriate for the non-expert user applying the model to obtain a single prescriptive result. Outputting too many results could confuse the non-expert, and obscure the interpretation of the important information.

In contrast, the full documentation included with a model does not vary with the identity of the end-user or the way the model is used. All models should have document-ation including the following: a clear statement of the model objectives; a description of the structure, hypotheses, assumptions and boundary conditions underlying the model; a report of the mathematical formulae used; a fully-commented computer implementa-tion; and an explanation of how the model is run. However, some of this documentation may be hidden from the end-user, to avoid overloading the non-expert. The end-user will often only see documentation describing how to run the model through the user interface provided.

Descriptions of some generic types of model application are given in this chapter. These are just some examples of model applications; there are many others that could have been described. We use these few examples to demonstrate general principles.

The prudent application of a model requires answers to the five key questions: Who is the end-user? What will the model do? How can the application guard against input error? How can the application guard against misinterpretation of the results? What documentation is needed? By considering each of these questions in turn, you should be able to construct from your model a working application that is useful and meaningful to the target end-user.

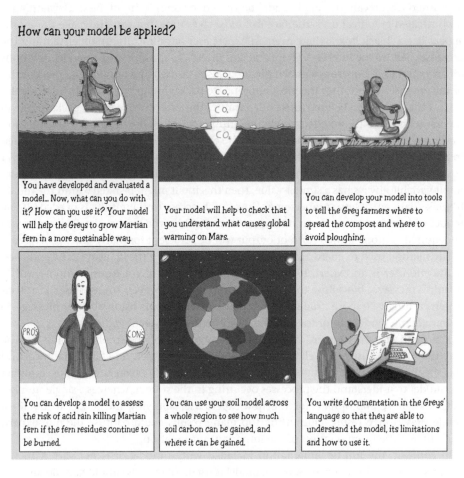

How can your model be applied?

You have developed and evaluated a model... Now, what can you do with it? How can you use it? Your model will help the Greys to grow Martian fern in a more sustainable way.

Your model will help to check that you understand what causes global warming on Mars.

You can develop your model into tools to tell the Grey farmers where to spread the compost and where to avoid ploughing.

You can develop a model to assess the risk of acid rain killing Martian fern if the fern residues continue to be burned.

You can use your soil model across a whole region to see how much soil carbon can be gained, and where it can be gained.

You write documentation in the Greys' language so that they are able to understand the model, its limitations and how to use it.

4.1 Scientific representation

Using a model for scientific representation can mean that the model is used without further packaging to interpret the experimental observations of a system and improve our understanding of the observations. Measurements of the way individual processes respond to changes in the system can be taken and used to develop the components of the model, but often it is only when these components are brought together (as already described in Chapters 2 and 3) that we can quantify how one process affects another. A model consisting of linked processes can sometimes provide more information about the controls on the individual processes than the direct measurements themselves.

The failure of a model that contains our best scientific knowledge, and that works well in other conditions, can greatly improve our understanding of the system. The failure of the model identifies missing knowledge and may even allow us to highlight which processes are misunderstood. People living around Aberdeen, Scotland, can often be heard to comment that the national weather forecasts '...never get it right for this corner'! The localised coastal effects in this protruding 'corner' of the British Isles mean that the mostly accurate models used by the Meteorological Office to provide weather forecasts for the rest of Britain very rarely get it right for Aberdeen! More accurate weather forecasts might only be achieved for Aberdeen by including these additional coastal processes. Knowing this tells you which processes you need to understand better, which sites the model is likely to fail at, and allows you to assess whether it is worthwhile to put more effort into correcting the model at these sites. Some cynical Aberdonians say that the London-based Meteorological Office have clearly decided not to put more effort into correcting the model for Aberdeen, but this is perhaps just local opinion!

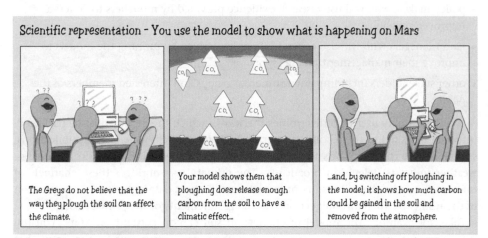

Scientific representation – You use the model to show what is happening on Mars

The Greys do not believe that the way they plough the soil can affect the climate.

Your model shows them that ploughing does release enough carbon from the soil to have a climatic effect...

...and, by switching off ploughing in the model, it shows how much carbon could be gained in the soil and removed from the atmosphere.

4.1.1 Who is the end-user?

There are two types of end-user for a model that provides a scientific representation of the system, namely, the primary and the secondary end-user. The primary end-user is the person who presses the keys on the computer—the computer operator who runs the model. The secondary end-user is someone who makes use of the information provided by the model, but does not actually run the model.

Primary end-users (who run the model themselves) could be any of the following:

- the scientist who developed the model, and so has complete knowledge of the model functions and access to the software used to create the model;

- the scientists who provided the measurements and information used to develop the model, and so may have knowledge of the model functions, but might not be able to directly access the model software;

- another scientist who was not involved in the model development, and so only has knowledge of the model functions provided by the developers, and often does not have access to the model software; or

- a non-expert who is using the model through a user interface to address specific questions.

Secondary end-users (who use the information but do not necessarily run the model themselves) could be any of the following:

- the scientist who developed the model, and will use the results to improve the representation of the system and learn more about the way the system works;

- the scientist who provided the information to develop the model, and will use the results to guide future areas of research (look for missing processes), justify continued research, and learn more about the way the system works;

- other scientists who will use information provided through published articles to obtain a greater understanding of the controls in the system;

- policy makers who will use research evidence provided by modellers to generate improved policies;

- land managers who will use general principles derived by running the models to improve their management practices; or

- other stakeholders including environmental groups and interested members of the public.

It is the primary user who determines the way in which the model must be packaged. Information is fed to the secondary user through media such as scientific publications, spoken advice (presentations to policy makers), leaflets, websites or (as in the case of weather forecasts) television broadcasts. The primary user abridges these channels of information from the full results of model runs. Correct data entry and interpretation of the results are therefore subject only to the requirements of the primary user.

Take, for example, the model of changes in soil carbon content on Mars. You use it to represent the effects of tillage on soil carbon loss. This persuades the Greys that tillage could be a major contributor to their problem. You, the scientist who developed the model, are the primary end-user. The Grey policy makers, who need to improve policies to encourage better soil management, are the secondary end-users. The Grey land managers do not need to run the model themselves, as you will interpret the model runs for them. It is *your* requirements that guide the packaging of the model. By identifying yourself as the primary user, it means that, for this purpose, you can save time by not packaging the model so that it can be used directly by the Greys.

4.1.2 How is the model used?

When a model is used to provide a scientific representation of the system, it can be used to:

- quantify the relative importance of the processes,

- interpret changes in the observations,

- or identify when the model fails, thereby identifying where our knowledge of the system is incomplete.

The relative importance of the processes can be quantified for the past, present or future representations.

The quantification of the relative importance of the processes in the past might be used to determine the extent or rate of the process required to achieve the present-day measured value. This use of models to explain how current measured values came about, sometimes known as reverse modelling, is an extremely useful device that provides much more information than is available from the measurement alone. For example, a process-based model of carbon turnover might be used to estimate the average carbon inputs from the Martian fern needed to achieve the present-day level of soil carbon as measured in the Martian fields. This information could not be obtained by measurement alone.

Quantifying the relative importance of the processes in the present day allows the contribution of each process in determining the current measured value to be deciphered. The soil carbon content measured on Mars may be low, but is that due to the impact of soil tillage, low plant inputs, or an inherent rapid turnover associated with the climatic and soil conditions? The Greys will be able to see how reduced tillage might affect the soil carbon content by reducing the amount of tillage included in the simulation. Is this sufficient to have a significant effect on the soil carbon stocks? The model application will be able to provide evidence to answer this question.

Quantifying the relative importance of the processes in the future allows the user to assess the effect of predicted changes in the values used to drive the model. If climate continues to change at its current rate, will the soil carbon content decrease further? What impact will changes in tillage have on the soil carbon content simulated under a changing climate? Is there any way of militating against these changes? Again, the model application can be designed to provide answers to these questions.

The interpretation of the observed changes in the field requires the model to be used to determine what controls the observed changes, and what feedbacks, discontinuities or points of no return exist. This information can be provided by a procedure similar to the sensitivity analysis described in Section 3.4. First, you must adjust the value of the parameters used to drive the equations determining the soil carbon content on Mars. The sensitivity of the soil carbon content to each parameter in turn must then be analysed; the controlling processes are identified with those equations to which the soil carbon content shows the highest sensitivity. The sensitivity of individual processes to each parameter in turn is then analysed; feedbacks can be identified for processes that show high sensitivity. Any discontinuities or points of no return will show up in the plot of the results of the sensitivity analysis as steps in the plot or an exponential change in the value.

Identifying where the model fails tells us which processes are not adequately modelled, which sites the model should not be used for, and allows us to assess whether it is worth doing more research to improve the model.

In the scientific representation of the soil carbon content on Mars, the model is used to quantify the amount of carbon loss attributable to tillage in the present day. Simulations using the future climate allow you to show the Greys how the soil carbon content is likely to change in the future if their tillage practices are left unchanged. You provide, for comparison, a simulation of how soil carbon content is likely to change in the future if tillage practices are improved.

4.1.3 Guard against input error

How the packaging of the model should guard against input errors depends on who the primary end-user is. If the primary end-user is the scientist who developed the model, then the checks for input error are a matter of personal preference, and will often be confined to the clear formatting of input files (if used), and warning and error messages from within the model when unusual inputs are encountered. The labelling of data within input files, using text that is unread by the model, is also a good idea.

If, however, the primary end-user is a scientist who was not directly involved in the development of the model, greater protection against input error is needed. The extent of the protective measures used depends on the capabilities of the end-user. If the end-user is the scientific community in general, then full measures should be used to guard against error in the inputs. A file describing the data types, giving the precise definition of the inputs and the units used, should be included in the model documentation. The warning and error messages should be more explicit, providing advice as to the range of input values allowed. A limited user interface might be required to help the scientist to enter data and run the model correctly. Default values could be provided, but at a minimum, a sample input file should be supplied so that the user knows what should happen when the model runs without error.

In the scientific representation of the soil carbon content on Mars, you set up your model with short warning and error messages to tell you if you have made a typing error, and you label the data in the input files. Since you are the end-user, no further protection against input error is needed, and so you waste no further time on this task.

4.1.4 Guard against misinterpretation of the results

How the packaging of the model should guard against the misinterpretation of the results depends not only on the primary end-user, but also on the use of the model results by the secondary end-user. The outputs from a quantitative model are numbers; how those numbers are presented can make the difference between a persuasive meaningful result and one that is hard to decipher. However, if the scientist working with the model has full access to the results as numbers, then spreadsheets, statistical packages and other interpretive software can be used to fully interpret the results. Often-used forms of graphical presentation may also be included in a simple user interface if desired, but it is the access to the numbers that is of paramount importance to the experienced scientist.

In the scientific representation of the soil carbon content on Mars, you use numbers produced during the simulations of future soil carbon content using current and improved tillage practices. You present these results in a form that is accessible to the secondary end-user, the Greys. You plot the predicted loss of soil carbon against time, both with and without improved tillage practices. Your presentation is well received and persuades them of the importance of changing their tillage policy.

4.1.5 Documentation

The model documentation should include the following: a clear statement of the model objectives; a description of the structure, hypotheses, assumptions and boundary conditions underlying the model; a report of the mathematical formulae used; a fully-commented computer implementation; and an explanation of how the model is run, including a description of the input data types, definitions and units used. For the purposes of scientific representation, all of this documentation should be freely available to the end-user, and the scientific rationale behind the model should be presented with the results.

In the scientific representation of the soil carbon content on Mars, you prepare a presentation for the Greys that includes the description of the structure, hypotheses, assumptions and boundary conditions underlying the model, together with information about where the model has been evaluated and how well it performed. This was sufficient to convince the Greys that the results you presented were reliable.

SELF-CHECK QUESTIONS: SCIENTIFIC REPRESENTATION

1. You develop a model to show the Greys what impact ploughing has on soil carbon and the potential impact this has on atmospheric carbon dioxide levels.

   Q: You will use the model and interpret the results. Which of the following measures are needed to allow you to use the model effectively?
   a. Give error and warning messages
   b. Give usual input value ranges
   c. Provide data labelling in the input files
   d. Provide a graphical user interface
   e. Limit outputs to single values
   f. Provide a user manual
   g. Provide full technical documentation
   h. Provide a help system in the user interface

   [A: All except e could be used, but since you, the developer, are the end-user, you might only *need* a and c.]

2. You now wish to develop a version of the model that will allow the Grey scientists to use the model themselves.

   Q: Which of the following measures are now needed to allow the Grey scientists to use the model effectively?
   a. Give error and warning messages
   b. Give usual input value ranges
   c. Provide data labelling in the input files
   d. Provide a graphical user interface
   e. Limit outputs to single values

    f. Provide a user manual

    g. Provide full technical documentation

    h. Provide a help system in the user interface

[A: Since the end-users are Grey scientists who did not develop the model, they will not know the idiosyncrasies of your model. You will therefore have to give them more help. They need to have full access to the outputs, so providing a single output (or a very limited number of outputs), as in e, would not be useful. All of the other measures could be useful, but at a minimum you should provide a, b, c, f and g.]

3. You now want a simplified version of the model to be loaded on the Martian Web so that Grey land managers and the Martian public can understand the problem.

    Q: Which of the following measures are now needed to allow the Grey land managers and the Martian public to use the model effectively?

    a. Give error and warning messages

    b. Give usual input value ranges

    c. Provide data labelling in the input files

    d. Provide a graphical user interface

    e. Limit outputs to single values

    f. Provide a user manual

    g. Provide full technical documentation

    h. Provide a help system in the user interface

[A: The Grey land managers and the Martian public will need a user-friendly interface to use the model, so you will definitely need to provide d and h. The error messages (a) and usual input value ranges (b) can be given within the user interface. Since the user will not see the input files, c is not necessary, but is still good practice to allow the interface developers to identify errors. In the case of these users, limiting the outputs to single outputs (e) on the Web-based application might help the user. The user manual (f) and all other documentation (g) should be available to the interested user.]

4.2 Expert and decision support systems

The model can be constructed into an expert system or a decision support system to help guide non-experts in decision processes that require expert knowledge. An expert system is a system that will give direct advice to the user, for example, 'Apply 50 kg N ha^{-1} on the 20th April'. A decision support system provides the user with information that supports the decisions they need to take, for example, 'If you apply 50 kg N ha^{-1} on the 20th April, 99% of the fertiliser will be taken up by the crop, but in 1 year out of 10, the crop will be limited by nitrogen availability.' Expert systems are generally easier to use than decision support systems, and require less interactive thought from the user. Decision support systems allow the user to make decisions based on their own priorities; is achieving the highest possible crop yield more important than reducing the pollution of the environment with leached nitrate, or is avoiding nitrate pollution in that river used for trout fishing of greater importance?

    Non-expert use of a model holds great potential for model misuse. In this case, it is the responsibility of the application developer to guard against all potential errors due to incorrect data entry and the misinterpretation of the results. Graphical user interfaces, warning and error messages, full sets of default data, the graphical presentation of selected results and easy-to-follow documentation on how to run the model are all essential tools in the application of models to provide expert advice and decision support.

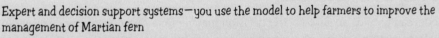

Expert and decision support systems—you use the model to help farmers to improve the management of Martian fern

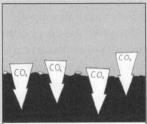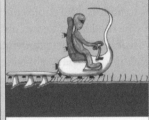

Your model tells you how much carbon can be locked up in the soil by changing the cultivation practice and by adding compost back to the soil.

The model can be built into a decision support system (with a user-friendly interface) to help farmers know where to avoid ploughing...

...and where and when to best apply the compost.

4.2.1 **Who is the end-user?**

In the application of a model for expert advice and decision support, the primary end-user (the computer operator who runs the model) and the secondary end-user (the person who makes use of the information provided by the model) are usually the same person, and that person is a non-expert. This does not imply that the user is necessarily a non-expert in the decision process for which advice and support are sought, but that the user *is* a non-expert in the use of the model, and perhaps also in the science underlying the model.

The end-user of an expert or decision support system might be a land manager, using the model to determine how best to manage their land at a particular site and in a given year. The general principles obtained from the results of field experiments or by running models for a range of different conditions are not sufficiently site-specific for this end-user. If a significant difference between the management recommendations is expected at different sites or in different years, then a model could be used as an aid to better decision making. The model will provide a virtual representation of the site, on which virtual field experiments can be run to determine the best course of action.

Back on Mars, your model can determine how much carbon can be locked up in the soil by changing the cultivation practice and by adding compost back to the soil. Also, because it is a predictive model, it can be used to look at future changes in soil carbon content. There is a significant difference in the effects on soil carbon of the timing of cultivation and composting for different soil types and in different years. If the model is built into an expert system or a decision support system, then it will help farmers to choose the best practice to increase soil carbon through changing cultivation and composting practice; so this is what you aim to do. The end-user is the farmer, and the system must be designed with the farmer in mind.

4.2.2 How is the model used?

A model can provide an aid to better decision making for a non-expert user by providing a single prescriptive recommendation (expert advice), or by presenting recommended options, with numbers or graphs to endorse the choice of options (decision support).

An expert system would usually be used if the system aimed to prescribe the priorities that the end-user should employ in obtaining a recommendation; if policy makers decide that reducing pollution is of higher priority than achieving the highest crop yield, then an expert system could be used to provide recommendations that are centred around this priority. The end-user would choose an expert system over a decision support system if there was insufficient time to collect and enter the more complex data needed to run a decision support system, or if the user was unable to decide between the recommendations proposed by the decision support system.

A decision support system would usually be used in preference to an expert system if the end-user needed to be convinced of the validity of the chosen course of action; presenting the scientific evidence that underlies the recommendation serves to persuade the user of the efficacy of the recommendation. The end-user would choose a decision support system over an expert system if recommendations are required that are tailored to the user's priorities, or if the nature of the land management was very different to that being employed by other managers.

Farmers on Mars are unwilling to adapt their cultivation and composting practices, which have remained unchanged for nearly a millennium. They are, however, very concerned about the loss of carbon from their soils, and the changing climate. There is no need to impose the priorities of the policy makers on the farmers; they share the same concerns but are unconvinced that it is worthwhile to learn new farming techniques. You decide to develop a decision support system from your model that will show them just how much they can increase the carbon content of their soils by changing their practices. The increases in carbon content are the results to be presented by the decision support system, and you should avoid confusing the farmers by presenting additional results that lie outside your objective. However, the system should also allow the farmers full flexibility to explore different options for carbon sequestration.

4.2.3 Guard against input error

The end-user of an expert or decision support system is usually a land manager—a non-expert in the application of the model, and possibly also the science underlying the model. This increases the potential for error in data entry. Errors in data entry occur due to mistakes in typing, mistakes in the formatting of the entered data, the wrong choice of units, mistakes in unit translation, misunderstanding of the underlying science, and mistakes in the running of the model.

Mistakes in typing can occur equally with the expert and non-expert user, and error messages should be used to catch any disallowed entered data. Warning messages should be used to catch unusually high or low data values, and, in an expert or decision support system, the user should be helped by informing them of the expected range of values.

Mistakes in formatting tend to occur when a user is unfamiliar with the rigours of accurate data entry—practice makes perfect! These mistakes occur most frequently when data is entered through a sequential input upon request by the model. Data entry through file input reduces these errors because it is possible to look back at the entered data and check that the values are correct, but errors are still common for the inexperienced model user. A graphical user interface is perhaps the most effective way of guarding against formatting errors; the data value is entered, and the system immediately checks for errors in inputs. Therefore, an effective expert or decision support system will usually run through some form of graphical user interface. Graphical user interfaces can be developed using Windows-based software languages, such as Visual Basic or Visual C++ (see the 'Further reading' section and **Web links 4.1 and 4.2**).

The wrong choice of units can be avoided by clear presentation of the correct units on the graphical user interface. A description of all the data types, giving the precise definitions of the inputs and units, should also be included, both online in context-sensitive help and in the user documentation.

Mistakes in unit translation can occur when the user is forced to translate from the units used in their records. There is no reason for this to happen. A good decision support system should allow data entry in a selection of different units, and the task of translating between units should be left to the user interface.

Errors in data entry due to a misunderstanding of the science can be avoided by the provision of a full set of default values; if the user does not know the value of the input variable, the system should provide it. Context-sensitive help, describing each process included in the model, should also be included.

Mistakes occur in running the model if the user documentation is too extensive for the busy land manager to read through. A summarised user documentation should be provided, focusing on how to run the model and where it can be used. Further background information should only be given to the user on demand through the context-sensitive help. A hierarchical structure should be employed that protects against changes to input values by the inexperienced user; this will allow the user to quickly gain confidence in the system but access more information as they become more practised.

The design of the decision support system for use by farmers on Mars should make it easy for the farmers to understand what is going on. Grey farmers have a very different way of thinking from Earth scientists, so the first stage is to visit the farmers and determine how they want the system to be laid out. You do this through a structured interview using a questionnaire. You are not an experienced programmer, so you develop your decision support system using Visual Basic. You can quickly become familiar with this programming language and start to develop your decision support system within the week. You include facilities to guard against the range of errors common in data entry. Data entered include the initial soil carbon content, weather data, the soil type, the crop type, the sowing and harvest date, the date of tillage and the compost type, amount and date of incorporation. Soon, you have a system that allows the farmers to run the model to simulate the effect of the tillage practices and the use of compost on the amount of carbon in the soil.

4.2.4 Guard against misinterpretation of the results

The outputs from the model are numbers, but, in contrast to the scientist, the non-expert end-user will not usually want to work with long lists of numbers. The way the results are interpreted is determined by the way the model is used.

An expert system requires a single recommendation to be provided. This can be obtained directly from the model, or, if the model does not directly produce the single result, it can be obtained by selecting the optimum value from a series of model runs. The expert system is internally responsible for driving the runs and selecting the optimum, so all the user sees is a single recommendation.

A decision support system presents a range of recommendations, selected according to different user priorities. A graphical representation of the results should show how the options differ, and demonstrate the validity of the model by a comparison with entered measured data. Statistical facilities can also be included to demonstrate model validity, but this should be presented in a way that avoids confusing a user who may not be fluent in statistics; the interpretation of the statistic should also be presented in words to ensure that the meaning is clear. A single prescriptive recommendation (such as that presented in the expert system) can also be included to help the user to interpret the results.

Both expert and decision support systems require extra facilities in the help system to describe the meaning of the results, and help the interpretation of the results.

In your decision support system of the changes in soil carbon on Mars, you want to demonstrate how soil carbon content changes with changes in cultivation and composting practice. The system runs a series of internal simulations to determine the timing of tillage and the rate of compost application that results in the optimum soil carbon content. The simulations are run using the specified initial soil, weather and cropping conditions. The results are presented as plots showing the soil carbon content against the timing of cultivation, and the soil carbon content against the amount and timing of compost incorporated. The plots indicate the recommended rate, and clicking on any point on the curve with the mouse will bring up plots showing the dynamic simulation of soil carbon turnover. Selecting a button labelled 'Check the accuracy of the simulations' takes the user into a higher level of the decision support system, where measurements can be entered and plotted against simulations. A statistical analysis of the simulations against the measurements is also presented here.

4.2.5 Documentation

As always, the model documentation should include the following: a clear statement of the model objectives; a description of the structure, hypotheses, assumptions and boundary conditions underlying the model; a report of the mathematical formulae used; a fully-commented computer implementation; and an explanation of how the model is run. However, for non-expert use of the model for expert advice or decision support, documentation should be limited to a summary of the structure of the model, a statement of where the model has been evaluated and a complete step-by-step guide to model use. Further documentation should be available to the user, but only on demand through the help system included with the model. Including the full documentation in the user manual is not recommended as it can be off-putting to the user and may result in the user not reading any of the manual at all!

The documentation of your decision support system of soil carbon content on Mars includes only installation instructions, a brief description of the model structure, a statement of where the model has been tested and shown to be reliable, and a step-by-step guide to using the model.

SELF-CHECK QUESTIONS: EXPERT AND DECISION SUPPORT SYSTEMS

1. You are trying to decide whether to develop your model into an expert system or a decision support system for use on Mars, to help Grey land managers to decide how much compost to apply and when to apply it to the fields.

   Q: Which of the following is typical of an output from an expert system, and which from a decision support system?
   a. Apply 20 tonnes of compost per hectare on sandy soils in spring
   b. The application of 10, 15 and 20 tonnes of compost per hectare on sandy soils in spring will cause soil carbon to increase by 0.1, 0.12 and 0.13 tonnes of carbon per hectare per year, respectively

   [A: Option a is typical of an expert system (it tells you what you should do). Option b is typical of a decision support system (it provides some information, but you then decide what is the best thing to do).]

   Q: Which of the following would be the best tool for educating the land managers about the importance of soil management?
   a. an expert system
   b. a decision support system

   [A: b. A decision support system does not prescribe a course of action, as an expert system often does, but allows the user to explore possibilities. This has greater potential to improve the user's understanding of the system and of the potential consequences of a range of options.]

2. Mars has an entirely different system of units to Earth (weight, area, etc. all have different units).

   Q: How would you ensure that the Grey land managers will enter the correct values into the decision support system you develop for them?
   a. Rewrite the model using their units
   b. Provide the user with a conversion table for converting their units into Earth units
   c. Make the user learn how to do the conversion for themselves without help
   d. Use the user interface to allow the user to select the units themselves, perform the conversions within the interface and allow the outputs to be viewed in different units

   [A: a, b and d are possible. c is not a useful option and will lead to the low use of the decision support system and frequent errors. Similarly, option b might lead to user error, so options a or d are preferable. Option d provides the most flexibility, but if the model is only to be used on Mars by Grey land managers then converting the units within the model is an option.]

4.3 **Risk assessment**

Models are used on Earth to assess the risks associated with a particular course of action. These can be related to human health, economics or environmental impact. Models are used throughout these risk assessment processes. Take, for example, environmental risk assessment, in which the risk of a chemical or other substance causing environmental damage is assessed. When assessing risk, there are two components that need to be considered, namely, the fate and distribution of the substance in the environment, and the likely impact of the substance on organisms or the environment.

For estimating the likely fate and distribution of a substance in the environment, a range of models can be used, often simulating the dispersion of the substance in the environment and in some cases estimating how much will end up in the atmosphere, water courses, soils, etc. The type of model used depends upon the substance being considered (e.g., whether it is a solid, liquid or gas), where it will be released into the environment (e.g., a point source like a factory or power station versus widespread sources such as domestic cleaning products) and its mobility (e.g., whether it disperses quickly, whether it all becomes dissolved, etc.). The fate and distribution models might also be used to assess the probability of given concentrations being exceeded, in a process using many model runs with slightly different inputs or assumptions, in a similar way to that described for sensitivity and uncertainty analyses in Sections 3.4 and 3.5.

For estimating the potential environmental impact and setting the standard for safe levels of a substance in the environment, the toxicity to indicator organisms (plants, invertebrates, fish, birds, mammals and whole microcosms) is assessed. This is often done in the laboratory, but models are often used here. Models predicting the effect of new chemicals from their structure, and from the known impact of similar chemicals, can be used. These are often referred to as QSARs (quantitative structural activity relationships). As the laboratory tests or QSARs have some uncertainty associated with them (e.g., other species of fish may be more sensitive than the one tested), safety factors are used. For example, the level known to be toxic to fish might be divided by a factor of 1000 or more for setting the safe concentration in the environment, sometimes called a predicted no effect concentration (PNEC) or similar term.

The risk assessment then involves comparing the expected concentration in given parts of the environment, determined by the fate and distribution models, with the environmentally safe levels determined with toxicity studies, QSARs and safety factors. Risks posed by that substance to particular organisms or environments, in particular parts of the environment, can therefore be assessed. If the probabilities of exceeding certain concentrations have been calculated by the fate and distribution models, then a risk probability can be calculated. Whether a given risk is acceptable or not is determined by society through, for example, consultation between regulators, industry, environ-mental groups and the public. A simplified scheme showing the main components of environmental risk assessment is shown in Fig. 4.1.

There are many risk assessment schemes available, but the US-EPA (United States Environment Protection Agency) has a good website explaining how it assesses the environmental risk posed by substances in the environment (**Web link 4.3**).

Fate and distribution in the environment

Environmental impact

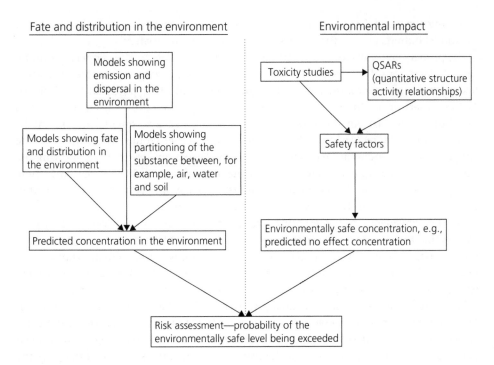

Fate and distribution in the environment:

Models showing emission and dispersal in the environment

Models showing fate and distribution in the environment

Models showing partitioning of the substance between, for example, air, water and soil

Predicted concentration in the environment

Environmental impact:

Toxicity studies

QSARs (quantitative structure activity relationships)

Safety factors

Environmentally safe concentration, e.g., predicted no effect concentration

Risk assessment—probability of the environmentally safe level being exceeded

Figure 4.1 A simplified scheme for environmental risk assessment.

Risk assessment—...but what are the risks of this happening?

Burning Martian fern residues led to acid rain on Mars. A risk assessment might have helped to predict that problem.

The acidity of the rainfall could have been measured and experiments could have been used to determine the toxicity of the acid rain to Martian fern.

Models describing the fate and distribution of the sulphur-containing pollutants in the atmosphere could have been used to estimate the acidity of the falling rain and the risk of burning Martian fern residue assessed.

4.3.1 Who is the end-user?

Since a range of models are used in risk assessment, at different stages in the process, there are multiple end-users. However, since it is important that all interested parties, including the general public, can see how the risk assessment has been derived, all of the model outputs used in the risk assessments must be transparent.

The primary end-users for the overall risk assessment models are the people performing the risk assessment. These results are then passed to the secondary users such as regulating bodies, industry, environmental groups, the public and other potential stakeholders. The end-users of the intermediate model outputs (e.g., the outputs from the fate and distribution models) are the people conducting the risk assessment, who might be from industry, the regulating body or from environmental groups. These intermediate end-users might require more detailed technical information, for example, showing the expected concentrations in water, soil and the atmosphere, and the variation in time and space, compared to the secondary end-users, who may simply wish to know whether or not a substance is likely to pose a risk to the environment.

Another consideration in presenting a transparent risk assessment to secondary end-users is that the models used vary in complexity. It is easier for a non-specialist to understand an environmentally safe level of a substance in the environment than to understand how a QSAR model estimates toxicity to non-target organisms. **Web link 4.3** provides a good example of how model outputs can be packaged to inform the non-expert end-user.

In assessing the risk of acid rain damage to Martian fern due to fern residue burning on Mars, the end-user of the overall risk assessment is the Martian policy maker, and the risk assessment system is designed with that user in mind. The end-user of the more detailed fate and distribution models used in the risk assessment is the scientist working for the regulating body or environment agency on Mars; the outputs for these intermediate users will be quite different from the overall risk assessment for the policy maker.

4.3.2 **How is the model used?**

Again, the range of models used in risk assessment presents a range of different modes of application. The policy maker, or a member of the public using a web-based version of the risk assessment tool (i.e., when acting as a primary end-user), might only see the environmental safe limit and the predicted concentration in air, water and soil in a given region. The data used in those applications will be derived from models, but will be numbers held in a database (sometimes called a look-up table) without the models being run each time a user accesses the risk assessment tool.

During the process of assessing the risk posed by a given substance, the expected concentrations in water, soil and the atmosphere, and the variation in time and space might need to be known by the person assessing the risk. In this case, the model is used in a similar way to that described in Section 4.1, since this is effectively a scientific representation of the likely distribution of the substance in space and time. The outputs communicated to the overall risk assessment end-user will be very different to the databases of model outputs used in assessing the risk.

For assessing the risk of acid rain damage to Martian fern due to fern residue burning on Mars you might use a model to determine how much of the sulphurous acidifying material is released when Martian fern is burned and another model to estimate how this material is moved around by prevailing winds. The same model (or another climate model) might be used to estimate where that acid rain is likely to fall. Assuming that the environmentally safe (or damaging) levels have also been determined, the risk of acid rain damaging Martian fern in different regions can be assessed.

4.3.3 **Guard against input error**

The end-user of a web-based risk assessment scheme is using a system similar to an expert system, described in Section 4.2. Similar techniques, such as warning messages, listing the ranges of expected inputs, etc., would be used in such a system. Graphical user interfaces can be developed using Windows-based software languages, such as Visual Basic or Visual C++ (see the 'Further reading' section and **Web links 4.1 and 4.2**).

For the detailed models used in making the risk assessment, the problems associated with input error are similar to those presented when using a model for scientific representation. This was discussed in detail in Section 4.1.3.

On Mars, for the scheme used to assess the risk of acid rain damage to Martian fern due to fern residue burning, a range of models will be needed. You will get your team to work on these different models. You ask your team to set up the models with short warning and error messages to tell other users in the team if they have made a typing error, and you label the data in the input files. Since the end-user for the detailed models is your team, no further protection against input error is needed. However, the web-based tool you use to allow Martians to assess risk in their region uses model outputs from these detailed runs (in a database). For this interface, you include facilities to guard against the range of errors common in data entry, as described in Section 4.2.3.

4.3.4 **Guard against misinterpretation of the results**

If the end-user is a non-expert, simply accessing the risk assessment outputs, then the outputs need to be simple. This might consist of an environmentally safe concentration in air, water or soil, as appropriate, a maximum expected concentration in these media, and perhaps a categorisation of risk on a quantitative (e.g., 1–10) or semi-quantitative (e.g., very low–low–medium–high–very high) scale. The outputs from the more detailed models might be more complex, giving predicted concentrations of a substance in different media at different locations over time. The same danger of potential misinterpretation of the results occurs as when using models for scientific representation (see Section 4.1.4).

In your scheme for assessing the risk of acid rain damage to Martian fern due to fern residue burning on Mars, the web-based user interface of the overall risk assessment should be relatively trouble free. However, you must make sure you give guidance on how to interpret qualitative statements of risk if you use them. One person's interpretation of 'low risk' will be very different from another's. Qualifying these statements with a probability, for example, 'low risk means there is less than 0.1% risk of Martian fern residue burning in the area causing damage to Martian fern', will help to avoid misinterpretation. For the application of the more detailed models, good documentation of the outputs and units used, as well as frequent team meetings, will help to avoid members of your team misinterpreting the results.

4.3.5 Documentation

For the detailed models used in making a risk assessment, the model documentation requirements are the same as for a model used for scientific representation (see Section 4.1.5). The web-based interface needs to be accompanied by clear statements about the way the risk assessment is derived, the assumptions made (e.g., model assumptions and any safety factors used), and statements to allow qualitative statements to be interpreted, such as an assignment of probability to risk, as described in Section 4.3.4. More technical documentation should also be accessible to the more interested user to ensure full transparency.

On Mars, it is important to gain the trust of the Greys that the system is completely transparent. You are, after all, a recently arrived alien race. The documentation of your risk assessment system on Mars includes a web-based tutorial, a step-by-step guide to using the model, and all of the manuals and supporting documents described above. All of the documents and reports of previous applications of the model are made available to the users.

SELF-CHECK QUESTIONS: RISK ASSESSMENT

1. In environmental risk assessment, models are used at many stages in the process.

    Q: In which of the following parts of an environmental risk assessment can models be used?
    a. Determining the fate and distribution of a substance
    b. Establishing the partitioning of a substance between air, water and soil
    c. Determining the likely concentration of a substance in different parts of the environment
    d. Determining the likely toxicity of a substance to various organisms or ecosystems

    [A: Models are used at all stages. For d, QSARs are a type of model that can be used to do this.]

    Q: Which of the following concentrations are compared in arriving at a risk assessment for a substance?
    a. The concentration of the substance in fish tissue
    b. The environmentally safe concentration of a substance
    c. The concentration of a substance that is toxic to plants
    d. The predicted concentration of a substance in various parts of the environment

    [A: b and d are compared (see Fig. 4.1). The other measures might be used along the way, but the overall risk assessment relies on comparing the expected concentration in the environment with the environmentally safe concentration.]

2. Non-expert users might find it easier to understand terms like 'low risk' compared to terms like '>0.1% probability of the environmentally safe concentration being exceeded'.

    Q: How do you make sure that the term 'low risk' is not misinterpreted by the user?
    a. Do not worry about it—this is the user's problem
    b. Provide a user manual that can be downloaded and that quantifies terms such as 'low risk' in terms of probability
    c. Provide a quantification of what the term means in terms of probability as part of the interface

d. Provide an online tutorial that allows the user to become familiar with the risk assessment tool

[A: c and d are the best ways to help avoid misinterpretation. b could be used, but few casual users will download and read the documentation so it cannot be relied upon. Option a will not help the user at all!]

4.4 Spatially-explicit applications

For assessing impacts at larger scales, models can be made spatially explicit by using geographical information systems to format the inputs and visualise the outputs. Spatially-explicit modelling is discussed in this section.

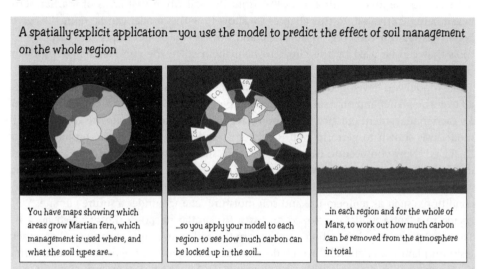

A spatially-explicit application—you use the model to predict the effect of soil management on the whole region

You have maps showing which areas grow Martian fern, which management is used where, and what the soil types are...

...so you apply your model to each region to see how much carbon can be locked up in the soil...

...in each region and for the whole of Mars, to work out how much carbon can be removed from the atmosphere in total.

4.4.1 Who is the end-user?

With spatially-explicit model applications, you are simply performing an application of a model similar to those described throughout the book. The only difference is that the model is run many times with different sets of inputs associated with a particular area of space, which is often done for scientific representation, but may also form part of decision support or expert systems, risk assessment schemes or a variety of other applications. The end-user will depend upon what the application is for. Let us assume in this section that you wish to perform a spatially-explicit application for improving your scientific understanding. In this case the primary end-users are yourself and other scientists. If the application is for policy-relevant outputs, then secondary end-users may include policy makers (who often like to see maps of the model outputs!) or the wider public. However, these secondary end-users do not run the model themselves, they simply use the packaged outputs as maps or spatial datasets. Given that you are the end-user, you can save time by not packaging the model so that it can be used directly by others.

In the example of soil carbon loss on Mars, the end-users are you and your team (many of whom are Greys). The model will not be used by the wider public.

4.4.2 **How is the model used?**

The model is used in a similar way to when it is applied at the site scale, except that it has to be run many times with different sets of inputs associated with a particular area of space, which might be a cell of a grid or a polygon representing an area of space. To describe how the model is applied, we will use our example from Mars. The example we give here addresses the questions 'How will soil carbon on Mars change under a future climate?' and 'How much CO_2 can be locked up in the soil by changing land management?' You have shown in Chapters 2 and 3 that climate change on Mars has been due to the loss of carbon (as CO_2) from the soil when vast areas of Martian fern began being cultivated. In this chapter we show how this model can be refined to give better answers for each region on Mars, and how it can be developed into a range of tools that can be used by the Greys to help them solve the climate change problem on Mars.

The model you developed in Chapter 2 was based on Martian soil data collected from all over the planet and on many different soil types. A single relationship was developed for each management practice that ignored regional differences in climate and soils. The model was shown to perform well (Chapter 3) and suggested that the problem could be solved. Now you need to develop a model to show how the effect differs among the different climate and soil types found in each region.

First of all, you need a model that tells you how soil carbon changes as environmental conditions, such as temperature and soil moisture, change under a future climate. You decide to adapt a model developed on Earth for similar purposes (RothC; **Web link 4.4**). This model assumes that carbon coming into the soil from plants is composed of either resistant plant material (RPM) which decomposes slowly (0.3 of the pool of carbon turns over each year), or decomposable plant material (DPM) which decomposes more quickly (the pool turns over ten times per year). Each time any pool of carbon decomposes, some of the carbon is lost to the atmosphere as carbon dioxide. The incoming plant material decomposes to soil microbial biomass (BIO), which itself decomposes quite quickly (0.66 of the pool turns over each year) to humic material (HUM) which is much more stable (0.02 of the pool turns over each year). Both BIO and HUM can be decomposed further to BIO and HUM with some carbon lost as carbon dioxide each time. Figure 4.2 shows the structure of the model. The rate of transfer of carbon between these pools is affected by environmental conditions such as temperature, soil moisture and plant cover.

To test the model you need the climate data to run it. You need monthly values for the total precipitation, total evapo-transpiration and mean temperature, the clay content of the soil (which stabilises the soil organic matter), estimates of the carbon inputs to the soil and estimates of the current soil carbon stock. You can get these for sites on Mars, so you test the model on many sites in many different regions covering many different soil types and climates, using the methods described in Chapter 3. The model is shown to work well.

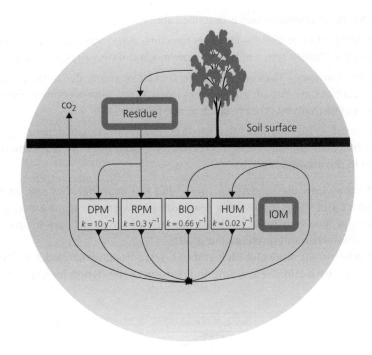

Figure 4.2 The Rothamsted soil carbon model (RothC).

You now need to run the model for the whole planet to answer the questions 'How will soil carbon change under a future climate?' and 'How much CO_2 can be locked up in the soil by changing land management?' For this, you need to get the same data to run the model, but you need data for the whole planet!

When you run models over a large spatial area you usually divide the surface into smaller units that are assumed to be homogeneous. These can be on a regular grid, or can be based on polygons (soil units or administrative regions, for example). The data held within the database for each polygon or grid cell is the same as the data you need to run the model at one site. In this case, the following are required:

1) a database of the current soil carbon stocks on Mars;

2) a database of the predicted climate in the future;

3) estimates of how Martian fern will respond to climate change and changing atmospheric CO_2 concentration, to estimate how carbon inputs to the soil will change; and

4) estimates of the extent to which each improved management practice can be implemented in the future.

To derive all of these sets of information that you will use with your soil model, you not only have to rely on measured data, but you also need outputs from other models, developed and tested by other modellers in the team. From experience of similar exercises on Earth (see Section 4.5), you will need the following.

1. *Soil carbon model (RothC)*—This is described above.

2. *Martian fern production model*—This uses the climate and soil data to predict what the growth of the Martian fern will be, and how much carbon will be returned to the soil.

3. *Soil transfer functions*—These are equations that allow you to derive the parameters that you do not have but are interested in (such as soil carbon content) from other soil parameters that you do have (such as soil type).

4. *Climate model*—This is a model that predicts changes in climate depending on (among other things) the concentration of greenhouse gases (such as carbon dioxide) in the atmosphere.

Figure 4.3 shows how the models you need, and the data that they use, fit together. You can plot all of these inputs on a map using a geographical information system. This shows you how your input variables that drive your model vary in space. This is extremely useful when interpreting your results.

You can now run the model for each grid cell. You automate the model so that it reads in the inputs for each grid cell at a time and performs a simulation for a specified period

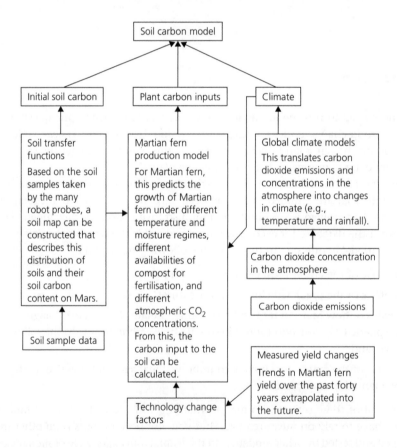

Figure 4.3 The outputs from other models and the data needed to run your soil model for all of Mars.

of time (e.g., 100 years), and then moves on to the next grid cell. The model will take a while to run and the output files will be large, but at the end it outputs the predicted soil carbon in each cell for each month or year. You can then analyse the outputs to look at changes in soil carbon over time (e.g., between now and 2050 or 2100). You can also plot maps of the model outputs to show how the change in soil carbon varies in different regions. You run the model, perform the analysis and show the following.

1. Increasing CO_2 concentrations in the atmosphere will increase soil temperatures and will speed decomposition, so more soil carbon will be lost as climate change occurs.
2. Warmer temperatures and increased CO_2 concentrations in the atmosphere will increase the growth of Martian fern and increase carbon inputs to the soil, partly counteracting the increased loss of carbon from the soil.
3. Halting tillage of the Martian fern fields will slow the loss of carbon from the soils.
4. Spreading compost on the field greatly enhances soil carbon.
5. Combining no-tillage with the utilisation of compost on the Martian fern fields stops carbon being lost from the soils to the atmosphere, gradually lowers atmospheric CO_2 concentrations, and slows climate change on Mars.

You can now put these management plans into action. Martian land managers need to start composting their waste (which will also help to stop the problem of acid rain) and avoid cultivating the soil of their Martian fern fields.

4.4.3 Guard against input error

Input error can occur very easily in spatial model applications for a number of reasons.

1. Due to the large volume of data, errors are inherently more difficult to spot.
2. At least some of the spatial datasets come from secondary sources, that is, data that you did not collect yourself. This means that:
   a. you have no control over data quality,
   b. the format may be confusing (e.g., many datasets multiply values with one decimal place in the data by 10 to remove the decimal place—i.e., a mean monthly temperature of 1.5 °C will be stored as 15 in the database—and thereby reduce the size of the file),
   c. the data may be derived from model outputs and the uncertainty associated with the values may be unknown, and
   d. the units may be different from those used by your model.
3. There are likely to be missing values for some input parameters in some cells within the grid. The missing value code may differ among datasets.
4. There are a variety of geographical projections used, and there are different conventions as to where the grid is measured from (e.g., the centre of the grid cell or the bottom left corner of the grid cell).
5. Data for different spatial parameters are often available at different spatial scales, and using different spatial formats (e.g., some are in grid format, and some in polygons).

To guard against input error, data must first be converted to a compatible projection and the grids cells or polygons overlaid. This is most easily accomplished within a geographical information system. Plotting the values on a map allows outliers to be spotted. Careful checks on the units used and on any multiplication factors need to be made. Where possible, the uncertainty in the input data should be quantified. This is essential when interpreting the results (see Section 4.4.4). All missing values need to be converted to a standard code that is recognised by the model.

In the Mars example, some of the data were derived from humans working on Mars and some from data collected by the Greys. Harmonising the datasets used to drive the models and error checking took a very long time!

4.4.4 Guard against misinterpretation of the results

Spatial model applications also present a number of potential areas for misinterpretation. Since the models often use data derived from other sources (and other models), it is not always possible to control data quality and uncertainty. Your model may use the input data and provide outputs that look very convincing when plotted on a map. However, if the input data were unreliable, then so will be the model outputs. Further, it is very important to know how your model responds to the full range of inputs used across the area for which it is applied. Sensitivity and uncertainty analyses, as described in Sections 3.4 and 3.5, respectively, are essential in interpreting the results. Where possible, the uncertainty in the input data should also be quantified.

Where applying anything other than the simplest of models, interactions between processes also occur. This is problematic enough in interpreting results at the site scale, but it is even trickier when applying a model spatially. To aid interpretation, the model outputs can be examined for individual grids cells, and if necessary the model can be re-run varying the inputs in a similar way to sensitivity and uncertainty analyses, as described in Sections 3.4 and 3.5, respectively.

In the Mars example, to gain the confidence of the Greys in your model application, you fully describe all of the input uncertainties and assumptions and provide a full sensitivity and uncertainty analysis of the model. The maps you plot are made public so that you can gain feedback on your estimates.

4.4.5 Documentation

In a similar way to any model used for scientific representation, the model documentation should include the following: a clear statement of the model objectives; a description of the structure, hypotheses, assumptions and boundary conditions underlying the model; a report of the mathematical formulae used; a fully-commented computer implementation; and an explanation of how the model is run, including a description of the input data types, definitions and units used. All documentation should be freely available and the scientific rationale behind the model should be presented with the results.

When applying the model spatially on Mars, you prepare a presentation for the Greys that includes the description of the structure, hypotheses, assumptions and boundary conditions underlying the model, together with information about where the model has been evaluated and how well it performed. You also provide maps of your outputs so that everyone can comment on the model outputs and any locally-specific conditions that may have been overlooked can be included. The Greys are convinced by your results and implement the improved management practices that you recommend.

SELF-CHECK QUESTIONS: SPATIALLY-EXPLICIT MODEL APPLICATIONS

1) When using spatial model applications, particular problems are posed in comparison to applying a model at an individual site.

   Q: Which of the following factors are expected to make input errors more likely?
   a. A much larger volume of data
   b. Some of the datasets come from secondary sources (i.e., they were not collected by yourself)
   c. There are likely to be missing values for some parameters for some grids cells or polygons
   d. The geographical projections used for the spatial datasets may not be the same
   e. The spatial form of the data may not be the same (e.g., small grids, large grids, polygons, etc.)

   [A: All are true.]

2) At least some of the spatial datasets you use in spatial model applications come from secondary sources, that is, data that you did not collect yourself. This can present problems.

   Q: Which of the following can occur when using spatial datasets not collected by yourself?
   a. The units may be different from those that you need
   b. The uncertainty associated with the input data is unknown
   c. The data will inevitably be of poorer quality than data you have collected yourself
   d. The data quality cannot always be assessed
   e. The formatting may be misleading

   [A: All except c. The data quality may be excellent.]

3) The model uses many sets of input variables.

   Q: Why might plotting maps of input variables be useful?
   a. It makes attractive maps which are good for showing in presentations
   b. You can easily spot outlying values and investigate whether they are real or errors
   c. The spatial pattern of the inputs can help you to interpret the spatial pattern of your model results
   d. Policy makers only want to see maps and are not interested in whether or not the model is correct.

   [A: b and c are the best reasons for plotting maps of input variables, though they are also useful in presentations a. Whilst maps do make excellent visual aids in explaining results and regional differences to policy makers, the underlying science does of course matter, so answer d is not correct.]

   Q. Why would a sensitivity and uncertainty analysis of the model help in interpreting the spatial results?
   a. It shows how the model will react over the range of inputs that might be encountered across a large region
   b. You can identify where the model goes wrong

c. It allows you to see how the uncertainty in the inputs will affect the uncertainty in your outputs

[A: a and c are certainly true. b is also true if some of the input variables are outside the range for which the model was developed.]

4.5 Epilogue

4.5.1 How has life on Mars been improved?

You are a hero...

Your model showed where and how to apply improved land management...

...and the Greys implemented all of the recommendations from your model.

Your decision support system was used by Grey land managers to help them manage their land effectively.

Over the years, soil carbon levels increased, and carbon dioxide levels in the atmosphere decreased.

Climate change slowed and Martian fern began to grow well. The environment improved and the Greys flourished.

Humans and Greys live peacefully together and you are one of the greatest heroes on Mars.

4.5.2 The real-Earth applications of the models used in this book

Hopefully you have found the Martian setting for the model development, testing and application to be fun. We chose to set the examples on Mars, not just because we like stories about aliens, but because it allowed us to demonstrate aspects of modelling without the complications of the real world! Nevertheless, most of the examples used in this book come from real-life models, developed on Earth. In this section, we describe the real-life applications that we used as the basis for some of our simplified Martian examples.

Soil carbon change under reduced tillage and the application of organic waste

In Chapter 2 we developed a simple model that was used to describe soil carbon changes under different types of agricultural management. In our example (Table 2.1 in Section 2.3.4), we list data from fourteen paired plots on Mars showing the effects of cultivation on soil carbon change. In reality, these are data from fourteen European long-term experiments examining the effects of tillage on soil carbon. The values used are the real data from real experiments. We originally used these data to estimate the impact of zero tillage on European soil carbon stocks, as described in Smith *et al.* (1998). Similarly, the data presented in Table 2.2 and Fig. 2.3 are from long-term field experiments in Europe. These data were used to estimate the effect of applying animal manure to croplands in Europe to increase soil carbon stocks, as described in Smith *et al.* (1997*a*). In another paper (Smith *et al.*, 2000), we reworked these (and other data on soil carbon changes under different agricultural practices) to estimate the impact soil carbon storage could have on Europe's greenhouse gas emission reduction targets under the Kyoto Protocol (see **Web link 4.5**). We found, using exactly the types of simple model described in Chapter 2, that Europe could hit its Kyoto target if all of these measures were implemented, and even individual practices could, by themselves, make significant contributions (see Fig. 4.4). These measures were not all implemented, and we have since used a similar approach (Smith *et al.*, 2005*a*) to show that agricultural carbon storage will make little or no contribution towards meeting Europe's Kyoto targets by the end of the first Kyoto commitment period (in 2012).

The use of plots and quantitative methods to calculate the accuracy of simulations

In Chapter 3 we presented an example of how different plots can be used to assess the accuracy of a model simulation. Figure 3.5 in Section 3.2.1 is a real example of a model used to describe the reduction in slug feeding due to the predation of nematodes on slugs. The model revealed that the reduction in slug feeding could be described as a function of nematode concentration and exposure time. Further details are given in Wilson *et al.* (2004), where the simpler plot was used to present the data more clearly. In Section 3.3 we described the quantitative methods that can be used to compare models to measured data. We collected these methods together from other studies, and developed some others ourselves, to make the first quantitative comparison of dynamic models describing soil carbon dynamics using data from seven long-term experiments from around the world (Smith *et al.*, 1997*b*); see Fig. 4.5. All of the methods described in Section 3.3 were applied for this real-life application, and have since been used in many diverse modelling studies (the study had been cited in over 150 other journal publications in May 2006).

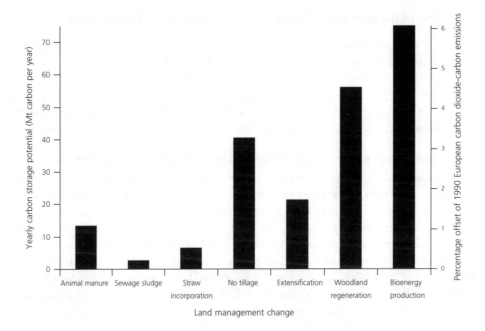

Figure 4.4 The carbon sequestration potential of cropland in Europe, showing the potential of different practices. The axis on the left shows the total amount of carbon that could be stored each year in millions of tonnes (Mt) of carbon. The axis on the right shows the percentage of Europe's Kyoto target that could be met by soil carbon storage. Europe's Kyoto reduction target is 8%, so some of these measures go a long way by themselves to meeting this target. More explanation about the practices used and how the estimates were derived is given in Smith et al. (2000)

The development of models into decision support systems

Earlier in this chapter, in Section 4.2, we described the application of a dynamic model in a decision support system. This is based on the development of the SUNDIAL model (Bradbury *et al.*, 1993; Smith *et al.*, 1996*a*) into a decision support system for fertiliser recommendation (SUNDIAL-FRS). SUNDIAL is a quantitative, deterministic, dynamic, functional and predictive model (see Chapter 1). It includes simple descriptions of all of the major processes of carbon and nitrogen turnover in the soil crop system (see Fig. 4.6).

The end-users of the system are farmers. The simple descriptions of the processes make SUNDIAL ideal for constructing into such a system, because it only uses data that is easily available on-farm. A Windows-based graphical user interface was devised to help the farmers run the model, and to guard against errors in the input and misinterpretation of the results.

So this was the planned system, but what exactly do farmers want the system to do? How do they want it to be laid out? What results do they want to see? To design the decision support system, a user group of 100 farmers was identified through their crop consultants. Each of the 100 farmers was visited, and structured interviews using a questionnaire allowed the requirements of the farmers to be ascertained (Smith *et al.*, 1996*b*).

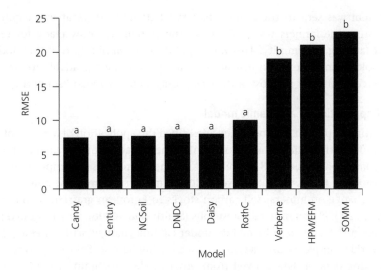

Figure 4.5 A comparison of the mean error (as indicated by the root mean squared error, RMSE) of nine different soil organic carbon models when used to simulate measured data from seven long-term experiments from around the world. The RMSE values of the models with the same letter (a or b) do not differ significantly (two-sample, two-tailed t test; $P >0.05$), but the RMSE values of the two groups (a and b) do differ significantly (two-sample, two-tailed t test; $P <0.05$). For more information on the models and datasets used, see Smith *et al.* (1997*b*).

In line with the farmers' requirements, the system was designed to provide a prescript-ive fertiliser recommendation (as would an expert system), but also allowed access to alternative recommendations and plots showing the flows of nitrogen to help persuade the user of the rationale behind the recommendation. The input values were checked for errors on data entry, and were supported by a complete list of default values.

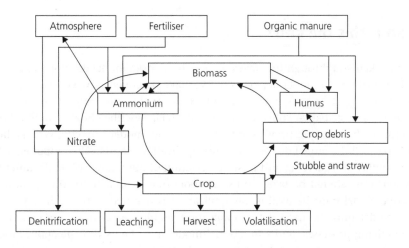

Figure 4.6 The processes of carbon and nitrogen turnover in the SUNDIAL model.

The system was sent to the farmers to test whether it did what was required. The feedback from the farmers was positive, so the system was now ready for release to the wider farming community. This was a 'public-good' application of the model, so it was not sold to make a profit; instead, the model was made available to farmers at a minimum cost (to cover the cost of the CD, postage, packaging and support).

Regional application of a dynamic model

Earlier in this chapter, in Section 4.4, we described the application of a model of soil carbon dynamics with spatial datasets to examine the impact of management, climate and land use change on Martian soil carbon stocks in the future. This is a simplified example of the application of the RothC model we made in Europe to examine the potential impacts of climate change on mineral soil carbon stocks in European grasslands and croplands (Smith *et al.*, 2005*b*) and mineral forest soils (Smith *et al.*, 2006), which formed part of a larger study on ecosystem vulnerability under climate change (Schröter *et al.*, 2005). In the real-world example we used a grid of data covering Europe. For each grid cell we used either existing data or data derived from other models, including the mean monthly temperature, precipitation, the potential evapo-transpiration for 1990–2080, the soil carbon content and percentage clay content from the European Soils Database, the land cover and land cover change, and the changes in the carbon input from plants and technology change. In the real-life application, the data and other models fitted together as shown in Fig. 4.7 (compare this with the simplified Mars application shown in Fig. 4.3).

The model was run using these data for each of the 31 000-plus grid cells to produce spatial outputs on the likely change in the mineral soil organic carbon each year between 1990 and 2080. The estimated change in the soil carbon in European croplands under the future climate by 2080, accounting for the direct climate impacts of soil, the effects on net primary production (plant growth) and the projected changes in technology, is shown in Fig. 4.8.

4.6 So is this the end?

We have taken you through the stages of modelling, from model development, through model evaluation, to model application. We have described how you can apply your model as a scientific representation, as an expert or decision support system, for risk assessment or in spatially-explicit modelling. We have described how you should answer the following five key questions in order to apply your model correctly: Who is the end-user? What will the model do? How can the application guard against input error? How can the application guard against misinterpretation of the results? What documentation is needed? Care should be taken to use a model that is suitable for the application. A predictive model must be used if the application is required to be predictive, a mechanistic model must be used if the application requires a scientific representation of the underlying processes, and a functional model must be used if the application is to produce a decision support system for use in the field.

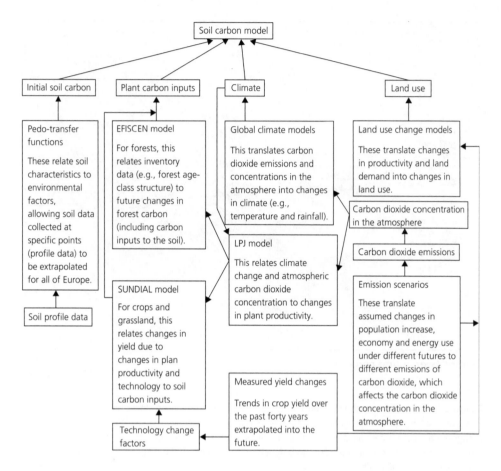

Figure 4.7 The outputs from other models and the data needed to run the soil model RothC to assess the likely changes in mineral soil carbon in Europe under climate change.

In this book, we have described the essentials of modelling, mainly from our own experience, but there is a whole world out there of different approaches, different ways of evaluating models and wonderful ways to make use of a model. Armed with the basic knowledge we have provided in this book, and confidence that it is not really as complicated as it might at first sound, you should be able to go on to explore this plethora of technologies and make use of the approaches that best suit your own purpose.

So is this the end? Have we done all that we can do with our model? Is the model finished? Perhaps there is no such thing as a completed model. For as long as the science continues to develop, and understanding of the systems around us is improved, so the models that summarise what we know about the systems will continue to develop. So this is not the end ... it is just the beginning! Happy modelling!

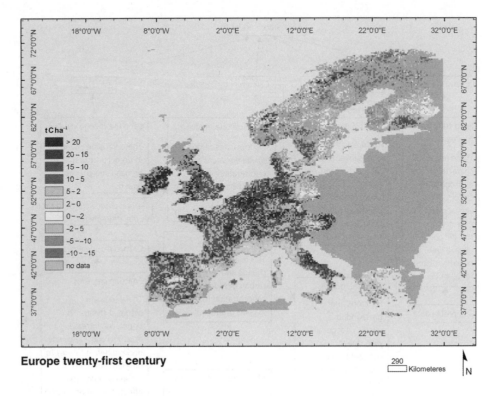

Europe twenty-first century

290 Kilometeres N

Figure 4.8 A map showing the difference in the mean soil organic carbon stocks between 1990 and 2080 for cropland for a single climate scenario (A2) as implemented by a single climate model (including changes in the net primary production and technology; HadCM3). The regional differences in the soil organic carbon stocks can be seen. For more details on the models and data used, and the assumptions made, see Smith *et al.* (2005*b*).

■ SUMMARY

1. Some examples of model application are given. These aid with
    a. scientific representation,
    b. expert advice or decision support,
    c. risk assessment, or
    d. regional-scale modelling.

2. The key questions to ask when applying a model are the following:
    a. Who is the end-user?
    b. What will the model do?
    c. How can the application guard against input error?
    d. How can the application guard against misinterpretation of the results?
    e. What documentation is needed?

3. Scientific representation
    a. The end-user could be

      i.   the scientist who developed the model,

      ii. the scientists who provided the measurements,

      iii. a scientist who was not involved in the development, or

      iv. a non-expert using the model through a user interface.

  b. Results will be used by

      i.   scientists,

      ii. policy makers,

      iii. land managers, or

      iv. other stakeholders.

  c. Models can be used to

      i.   quantify the relative importance of processes in the past, present or future,

      ii. interpret observed changes in terms of controls, feedbacks, discontinuities and points of no return, and

      iii. identify when the model fails—which processes and sites, and is it worth doing more research to improve the model?

  d. Guard against input error

      i.   If the end-user is the model developer then use the following: warning and error messages from within the model; and clear formatting of the input files including the labelling of data.

      ii. If the end-user is not the model developer then use the following: a file describing the data types, a definition of the inputs and the units used; more explicit warning messages from within the model; the development of a limited user interface to help with the input of data; and the provision of default values, at minimum an example input data file.

  e. Guard against misinterpretation of the results

      i.   Access to the results as numbers is essential.

      ii. The presentation should be tailored to the user of the results.

  f. Documentation

      i.   A clear statement of the model objectives.

      ii. A description of the structure, hypotheses, assumptions and boundary conditions underlying the model.

      iii. A report of the mathematical formulae used.

      iv. A fully-commented computer implementation.

      v. An explanation of how the model is run, and the results.

4. Expert and decision support systems

  a. The end-user could be

      i.   a land manager,

      ii. a non-expert in the use and science of the model, or

      iii. considering an issue where significant differences in recommendations are observed between years and sites.

  b. Models can be used for the following:

      i.   a prescriptive recommendation (expert advice)—used when the system aims to prescribe user priorities, the user feels unable to decide between the proposed options, or the user has insufficient time to run a decision support system;

      ii.   a recommendation that presents options (decision support)—used when the user needs to be convinced of recommendations, requires decisions tailored to their own priorities, or uses unusual management practices.

  c. Guard against input error

      i.   Mistakes in typing—use error and warning messages to catch disallowed data entry, and provide the expected range.

      ii.   Mistakes in formatting—the use of a graphical user interface for data entry will reduce errors.

      iii.   Wrong choice of data units—present the data units in the graphical user interface, and provide a description of the units (both online and in the user documentation).

      iv.   Mistakes in unit translation—allow data entry using a selection of units.

      v.   Misunderstanding of the science—provide a full set of default values, and context-sensitive help that describes the processes.

      vi.   Mistakes in running the model—provide summarised user documentation, and protect against changes to the input variables that are not required.

  d. Guard against misinterpretation of the results

      i.   Expert system—either a single recommendation obtained directly from the model, or a range of values resolved into a single optimum.

      ii.   Decision support system—presents a range of recommendations according to different user priorities, using graphical representation to demonstrate how options differ, and demonstrating the validity of the model using facilities to compare against the measured data.

      iii.   Both systems need a help system to describe the meaning and interpretation of the results.

  e. Documentation

      i.   A summary of the structure of the model should be available to the user.

      ii.   A statement of where the model has been evaluated should be provided.

      iii.   A complete step-by-step guide to model use should be available.

      iv.   Further documentation should be available to the user on demand through the help system.

5. Risk assessment

  a. Many models are used at many stages of the risk assessment process.

  b. The primary end-user for the overall risk assessment is the person performing the risk assessment and may be

      i.   a regulator,

      ii.   a scientist working in industry, or

      iii.   an environmental scientist.

  c. Results will be used by

      i.   regulators,

      ii.   industry,

      iii.   environmental groups, and

      iv.   the public and other stakeholders.

  d. Models can be used to

       i.   predict the likely toxicity to non-target organisms or ecosystems from the structure of the molecule,

      ii.   establish the partitioning of the substance to air, water and soil,

     iii.   determine the likely fate and distribution of a substance in the environment and its likely concentration,

     iv.   establish the probability or risk of an environmentally safe concentration of a substance being exceeded, and

      v.   show the public and other stakeholders how risk assessment is conducted.

  e.   Guard against input error

      i.   Depending on the type of model, the approaches adopted are similar to those used for scientific representation (for detailed fate and distribution modelling; see above) or for decision support (for Web-based tools for use by the public; see above).

  f.   Guard against misinterpretation of the results

      i.   Depending on the type of model, the approaches adopted are similar to those used for scientific representation (for detailed fate and distribution modelling; see above) or for Web-based tools for use by the public; the quantification of qualitative terms such as 'low risk' is essential.

  g.   Documentation

      i.   Depending on the type of model, the approaches adopted are similar to those used for scientific representation (for detailed fate and distribution modelling; see above) or for decision support (for Web-based tools for use by the public; see above).

6.   Spatially-explicit model applications

  a.   The end-user is usually the model developer, but could be

      i.   another scientist,

      ii.   a user of a spatially-explicit decision support or expert system (see above), or

     iii.   a user of a spatially-explicit risk assessment model (see above).

  b.   Models can be used to

      i.   do all the same things as any model can do when applied at the site scale, but for larger areas.

  c.   Input error spatial applications present a higher risk of input error due to

      i.   the large volume of data,

      ii.   some spatial datasets being from secondary sources,

     iii.   missing values for some input parameters in some cells,

     iv.   a variety of geographical projections being used, or

      v.   different spatial scales and using different spatial formats.

  d.   This can be guarded against through careful checking and the plotting of input values on a map.

  e.   Guard against misinterpretation of the results

      i.   Plot the input values on a map.

      ii.   Perform a sensitivity and uncertainty analysis.

     iii.   Examine the model behaviour at individual grid cells.

  f.   Documentation

      i.   The documentation is the same as for models used for scientific representation (see above).

7. The real-life application of the models used in this book
   a. Many of the models used in this book are based on real-life applications.
      i. The simple soil model used in Chapter 2 was based on a model to examine management impacts on soil carbon stocks in European croplands.
      ii. The slug–nematode example used in Chapter 3 to show the value of plots in the model evaluation was taken from a real model.
      iii. The methods described throughout Chapter 3 for the quantitative evaluation of models were collated and developed for comparing nine soil carbon models against data from long-term experiments. The method has since been used in many studies.
      iv. The decision support system used in this chapter is based on a real fertiliser recommendation system.
      v. The spatial application of the soil model on Mars was based on a real application to assess soil carbon changes under climate change in Europe.

■ PROBLEMS (SOLUTIONS ARE IN APPENDIX 1.4)

4.1. **You develop a scientific representation** of goat productivity in East African smallholdings. You will use this model to determine the factors controlling longevity, the production of young and lactation. This information will be used to prepare a scientific paper on the controls of goat longevity and productivity in East Africa.
   a. Who is the end-user?
   b. What will the application do?
   c. How should the application guard against input error?
   d. How can the application guard against misinterpretation of the results?
   e. What documentation is needed?

4.2. **You develop an expert system** to provide nitrogen fertiliser recommendations on UK farms.
   a. Who is the end-user?
   b. What will the model do?
   c. How can the application guard against input error?
   d. How can the application guard against misinterpretation of the results?
   e. What documentation is needed?

4.3. **You develop a decision support system** to provide pesticide application advice for use on tropical palm plantations.
   a. Who is the end-user?
   b. What will the model do?
   c. How can the application guard against input error?
   d. How can the application guard against misinterpretation of the results?
   e. What documentation is needed?

4.4. **You wish to develop a risk assessment** scheme for assessing the impacts of a new herbicide to get rid of weeds from Martian fern fields on Mars.
   a. Who is the end-user?

b. What will the model do?

c. What types of data might you need to set up the risk assessment scheme?

d. What types of model might you use as part of the risk assessment scheme?

e. How can the application guard against input error?

f. How can the application guard against misinterpretation of the results?

g. What documentation is needed?

4.5. **How would you develop and use a spatially-explicit application** of a model to estimate the effects of climate change on forest growth in the northern hemisphere?

a. Who is the end-user?

b. What will the model do?

c. How can the application guard against input error?

d. How can the application guard against misinterpretation of the results?

e. What documentation is needed?

■ FURTHER READING

Visual Basic

Foxall, J. (2003). *SAMS teach yourself Microsoft Visual Basic. NET 2003 in 24 Hours*. SAMS Publishing, Indianapolis, IN.
(*A straightforward, step-by-step approach to learn the essentials of Visual Basic. NET.*)

Wang, W. (1998). *Visual Basic 6 for dummies (for Windows)*. Wiley, Chichester.
(*A witty, well-written guide to Visual Basic 6.0.*)

Visual C++

Chapman, D. (1998). *SAMS teach yourself Visual C++ 6 in 21 days*. SAMS Publishing, Indianapolis, IN.
(*Covers all the essentials of basic Windows and Microsoft Foundation Classes (MFC) development, and addresses new features in Visual C++ 6.*)

Davis, S. R. (2000). *C++ for dummies* (4th edn, completely revised). Wiley, Chichester.
(*How to write programs, create source codes, use the Visual C++ help system, build objects, develop C++ pointers, debug programs, and more.*)

■ REFERENCES

Bradbury, N. J., Whitmore, A. P., Hart P. B. S. and Jenkinson, D. S. (1993). Modelling the fate of nitrogen in crop and soil in the years following application of ^{15}N-labelled fertilizer to winter wheat. *Journal of Agricultural Science, Cambridge*, **121**, 363–79.

Schröter, D., Cramer, W., Leemans, R., Prentice, I. C., Araújo, M. B., Arnell, N.W., *et al.* (2005). Ecosystem service supply and human vulnerability to global change in Europe. *Science*, **310**, 1333–7.

Smith, J. U., Dailey, A. G., Glendining, M. J., Bradbury, N. J., Addiscott, T. M., Smith, P. *et al.* (1996*a*). Constructing a nitrogen fertilizer recommendation system: What do farmers want? *Soil Use and Management*, **13**, 225–8.

Smith, J. U., Bradbury, N. J., and Addiscott, T. M. (1996*b*). SUNDIAL: A user friendly, PC based system for simulating nitrogen dynamics in arable land. *Agronomy Journal*, **88**, 38–43.

Smith, P., Powlson, D. S., Glendining, M. J. and Smith, J. U. (1997*a*). Potential for carbon sequestration in European soils: preliminary estimates for five scenarios using results from long-term experiments. *Global Change Biology*, **3**, 67–79.

Smith, P., Smith, J. U., Powlson, D. S., McGill, W. B., Arah, J. R. M., Chertor, O. G. *et al.* (1997*b*). A comparison of the performance of nine soil organic matter models using datasets from seven long-term experiments. *Geoderma*, **81**, 153–225.

Smith, P., Powlson, D. S., Glendining, M. J. and Smith, J. U. (1998). Preliminary estimates of the potential for carbon mitigation in European soils through no-till farming. *Global Change Biology*, **4**, 679–85.

Smith, P., Andrén, O., Karlsson, T., Perälä, P., Regina, K., Rounsevell, M. and van Wesemael, B. (2005*a*). Carbon sequestration potential in European croplands has been overestimated. *Global Change Biology*. **11**, 2153–63.

Smith, J. U., Smith, P., Wattenbach, M., Zaehle, S., Hiederer, R., Jones, R. J. A. *et al.* (2005*b*). Projected changes in mineral soil carbon of European croplands and grasslands, 1990–2080. *Global Change Biology*, **11**, 2141–52.

Smith, P., Smith, J. U., Wattenbach, M., Meyer, J., Lindner, M., Zaehle, S. *et al.* (2006). Projected changes in mineral soil carbon of European forests, 1990–2100. *Canadian Journal of Soil Science*, **86**, 159–69.

Wilson, M. J., Glen, D. M., Hamachar, G. M., and Smith, J. U. (2004). A model to optimise biological control of slugs using nematode parasites. *Applied Soil Ecology*, **26**, 179–91.

■ WEB LINKS

Web link 4.1: **http://msdn.microsoft.com/vbasic/**
Microsoft Visual Basic Developer Center

Web link 4.2: **http://msdn.microsoft.com/visualc/**
Microsoft Visual C++ Developer Center

Web link 4.3: **http://www.epa.gov/epahome/models.htm**
United States Environment Protection Agency (US-EPA) model pages

Web link 4.4: **http://www.rothamsted.bbsrc.ac.uk/aen/carbon/rothc.htm**
RothC home page

Web link 4.5: **http://unfccc.int/resource/docs/convkp/kpeng.html**
Kyoto Protocol on the UNFCCC website

Solutions to problems

1.1 Solutions to problems in Chapter 1

1.1. How would you classify the following model? A model has been developed to allow farmers to predict the potential increase in the weight of their cows with the amount of concentrate feed given each day. The model is structured around the following hypothesis:

> 'There is a 95% probability that the weight of cows will increase by 0.1–0.2 kg with each additional kilogram of feed given each day.'

Specify the type of
a. outputs,
   The output is the probability that the weight of cows increases by a given range. The output is quantitative and stochastic.
b. inputs,
   The inputs are 'kilograms of feed each day', and so can change with time through the simulation and are dynamic.
c. scope and
   The model must predict what will happen in the future, so it is predictive.
d. application.
   The model is to be used by farmers, so the application is functional.
   The type of model is quantitative, stochastic, dynamic, predictive and functional.

1.2. How would you classify the following model? A model has been developed to help policy makers estimate the changes in European soil carbon with changes in land management. This model must be used to quantify how soil carbon stocks might change in response to changes in land management and to determine improved management methods that might be used to increase soil carbon stocks. It is structured around the following hypotheses:

Hypothesis 1: 'Land management (manure management, tillage practice and crop residue management) determines soil carbon stocks in cropland',

Hypothesis 2: 'Changing land management will change soil carbon stocks to a level determined by the new management regime.'

Specify the type of
a. outputs,
   The output is the future soil carbon stocks. The output is quantitative and deterministic.
b. inputs,
   The inputs are the current soil carbon stocks and the type of land management. These do not change as the simulation proceeds, so the inputs are static.
c. scope and
   The model must predict what will happen in the future, so it is predictive.
d. application.
   The model is to be used by policy makers, so the application is functional.
   The model is quantitative, deterministic, static, predictive and functional.

1.3 How would you classify the following model? A model has been developed to help researchers estimate the size of a dolphin population, based on the probability of sighting a previously tagged dolphin on successive visits. The model is based on the following hypothesis:

> 'The probability of sighting a previously tagged dolphin is dependent on the size of the dolphin population.'

Specify the type of
a. outputs,
   The output is the size of the dolphin population. The output is quantitative and deterministic.

b. inputs,

The input is the probability of sighting a previously tagged dolphin, which can be determined by the number of sightings of previously tagged dolphins compared to the number of trips where the dolphins were not sighted. The input is static.

c. scope and

The model must estimate the size of the dolphin population at sites other than the site where the model was developed. The model is predictive with respect to location.

d. application.

The model is to be used by researchers to determine the population size. In the current application, they do not need to know how sightings and the population size are linked, so the model is functional.

The model is quantitative, deterministic, static, predictive and functional.

1.4. **How would you classify the following model?** A model has been developed to help bioenergy growers to determine which biofuel crops can be grown at different sites. The model is based on the following hypotheses:

Hypothesis 1: '*A biofuel crop will not grow outside a specified temperature range during the growing season*',

Hypothesis 2: '*A biofuel crop will not grow above a specified elevation*',

Hypothesis 3: '*A biofuel crop will not grow outside a specified rainfall range.*'

Specify the type of

a. outputs,

The output is the growth of the biofuel crop (yes or no). The output is qualitative.

b. inputs,

The inputs are the temperature range, the elevation and the rainfall range. The inputs

do not change during the model simulation, and so are static.

c. scope and

The model must determine whether biofuel crops will grow at different locations. It is predictive with respect to location.

d. application.

The model is to be used by bioenergy growers to determine which biofuel crops could be grown at a particular site. The application is functional.

The model is qualitative, static, predictive and functional.

1.2 Solutions to problems in Chapter 2

2.1. **Construct a conceptual model** of the time taken to get served in a queue for tickets at the railway station.

a. **List the reasons for developing the model.**

The purpose of the model is to determine in which queue I will get served most quickly. I should be able to use the model while standing in the railway station, from simple observations made in the railway station, without the assistance of a computer.

b. **Determine the type of model needed.**

Outputs: I need to know the time to get served, so the outputs are quantitative. The rule of parsimony suggests that a deterministic model is required.

Inputs: Single value inputs are needed. The inputs are static.

Scope: The model must predict the time to get served. It is predictive.

Application: I need to know how long it will take to get served. I am not interested in why it takes so long. The application is functional.

The type of model is quantitative, deterministic, static, predictive and functional.

c. **Draw a picture of the system.**

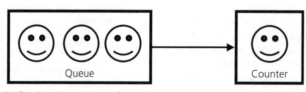

(Defined by the number of people)

d. List the hypotheses.

> **Hypotheses**
>
> 'The time taken to be served depends on the number of people in the queue and the average time taken for each person to be served.'

e. List the assumptions.

> **Assumptions**
>
> 1. 'The time taken to be served can be adequately represented by the average time taken.'
>
> 2. 'No additional people enter the queue in front of you after you have joined the back of the queue.'
>
> 3. 'Each person in the queue does a similar type of transaction (e.g., buys a ticket for a train).'

f. List the boundary conditions.

> **Boundary conditions**
>
> The model will adequately represent the system after 7a.m. and before 9a.m. After 9a.m. and before 7a.m., assumptions 1 and 3 would be expected to fail (e.g., as more people come to the station to buy railcards and tickets for journeys with complicated itineraries).

In the first queue, you observe that the average time to get served is about 60 seconds; in the second queue, this time is 90 seconds; while in the third queue, customers get served on average in 100 seconds. There are 10 people in the first queue, 7 in the second queue, but only 5 in the third queue. Construct a mathematical model of the three queues. Use a spreadsheet, a calculator or your head (!) to determine which queue you should join.

Mathematical model of the time to get served:

Time to get served = Number of people in queue × Average time for each person to get served

Queue 1: Time to get served = 10 people   × 70 seconds/person

= 700 seconds.

Queue 2: Time to get served = 7 people   × 90 seconds/person

= 630 seconds.

Queue 3: Time to get served = 6 people   × 120 seconds/person

= 720 seconds.

I should join queue 2, while remembering that the model has not been evaluated so I cannot assess how likely it is that I will get it wrong! (If you catch the same train every morning you could evaluate your model over the week and use it to choose the best queue in the following week.)

2.2. **Use statistical fitting procedures in a spreadsheet such as Microsoft Excel to derive a mathematical model** of the change in the height of trees with time after planting directly from the following data.

Time after planting (y)	Height of trees (m)	Time after planting (y)	Height of trees (m)	Time after planting (y)	Height of trees (m)
1	0.7	8	4.25	13.5	7.8
1.5	0.7	8.5	4	14	8
2	0.75	9	5.25	15	8.25
3	1	9.5	6	15.5	8.4
3.5	1	10	6	16	8.35
4	1.25	10.5	7	17	8.4
5	1.75	11	6.75	17.5	8.6
5.5	1.5	11.5	6.5	18	8.45
6	2.5	12	17.25	19	8.45
7	3.25	13	7.75	20	8.5
7.5	3.3				

Type the data into two columns of an Excel spreadsheet.

Select the data and use the 'Chart Wizard' (shown by the bar chart symbol on the toolbar) to plot the height of trees against the number of years. In the chart, select the data series for the height of trees by clicking on a point with the right mouse button. Select 'Add Trendline' and choose the type of trend line to be a polynomial of order 6 (trial and error shows that this trend line gives the best fit to the data). Under the 'Options' tab, choose to display the equation on the chart, and to display the R-squared value on the chart. The equation given is

$$\text{Height} = (-7 \times 10^{-6}) \text{ Years}^6 + 0.0005 \text{ Years}^5$$
$$-0.0113 \text{ Years}^4 + 0.1229 \text{ Years}^3$$
$$-0.5434 \text{ Years}^2 + 1.1135 \text{ Years}$$
$$-0.0412,$$

with an R^2 value of 0.9934 (the meaning of R^2 is explained in Chapter 3).

Other forms of equations could equally be used to achieve a high R^2 value.

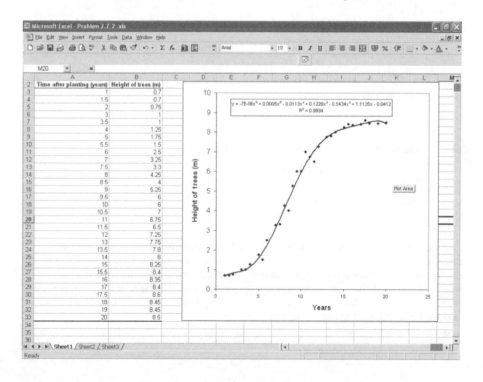

The model derivation in Excel is given on the website accompanying this book (**Web link 2.24**).

2.3. Carbon stocks on Mars change with the application of organic wastes according to the equation following:

$$
\begin{array}{ccc}
\textit{Future soil carbon} & & (0.0145 \times \textit{Amount of organic} \\
\textit{stock under organic} \ = \ \textit{Present soil carbon stock} \ + & \textit{waste} \times \textit{Years of} \\
\textit{waste application} & & \textit{organic waste management)} \\
\\
(\text{t C ha}^{-1}) & (\text{t C ha}^{-1}) & (\text{t C t DM}^{-1} \times \text{t DM ha}^{-1}\text{y}^{-1} \times \text{y})
\end{array}
$$

The following data is for use in the calculations.

	Present soil carbon stock (t C)	Arable area (ha)	Average application rate of organic manure for the sector (t DM ha^{-1} y^{-1})
Sector 1	43 460 000	820 000	50
Sector 2	134 408 000	2 536 000	15
Sector 3	618 828 000	11 676 000	25
Sector 4	121 741 000	2 297 000	40
Sector 5	793 993 000	14 981 000	10
Sector 6	955 696 000	18 032 000	0
Sector 7	39 962 000	754 000	20
Sector 8	472 601 000	8 917 000	2
Sector 9	3 021 000	57 000	7
Sector 10	48 866 000	922 000	8
Sector 11	74 253 000	1 401 000	7
Sector 12	117 236 000	2 212 000	2
Sector 13	136 740 000	2 580 000	3
Sector 14	147 340 000	2 780 000	50
Sector 15	322 028 000	6 076 000	25
Total	4 030 173 000	76 041 000	

Use the data in the table above to determine the total changes in carbon stocks over the next 10, 20, 30, ..., 100 years, assuming organic manure applications remain unchanged, and carbon stocks continue to increase with continued manure additions. Which sector will have the highest carbon stock after 50 years of manure applications? Which sector will have the highest carbon stock after 100 years of manure applications?

The completed model is given on the website accompanying this book in Microsoft Excel (Web link 2.12), Microsoft Visual Basic (Web link 2.21),

Compaq Visual Fortran (Web link 2.22) and Microsoft Visual C++ (Web link 2.23). You may find it useful to compare these different approaches in deciding which type of computer model you prefer to develop.

a. **In a general spreadsheet, such as Microsoft Excel,** construct the complete model of the carbon stocks of Martian land with organic waste application.

The input data used to drive the model are the present soil carbon stocks, the arable area of each sector, and the average application rate of organic manure. Enter these input variables into columns in Excel.

The number of years of manure application are also input data to the model. Enter the number of years in a row, to form a table area for the results.

The figure below shows the computer screen for this simple model in Excel. The formula bar shows the equation used to calculate the soil carbon stocks in each cell of the table area.

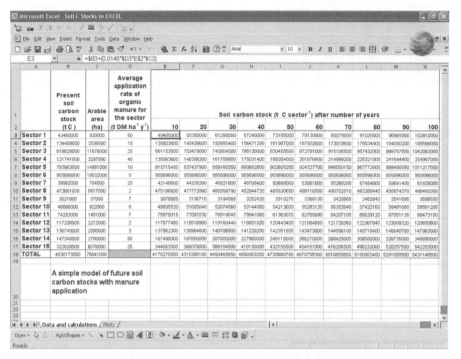

To help to understand the results, plot the soil carbon stock for each sector. The computer screen of the plot in Excel is shown below.

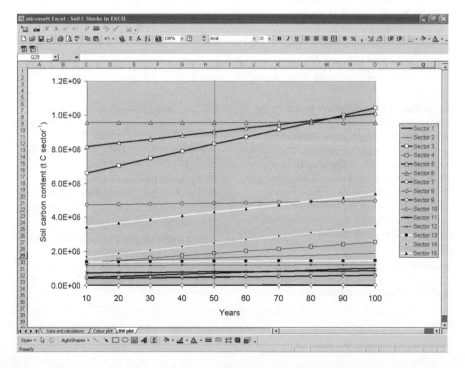

After 50 years of manure applications, sector 6 represents the highest carbon stock. After 100 years, the continued high manure applications in sector 3 are predicted to result in the highest carbon stock being in sector 3.

This simple model is provided in Microsoft Excel on the website accompanying this book (Web link 2.12).

b. **If you can program in Visual Basic**, construct the complete model of the carbon stocks of Martian land with organic waste application in Visual Basic.

The program in Microsoft Visual Basic 6.0 is given at Web link 2.21. The Visual Basic code is shown below.

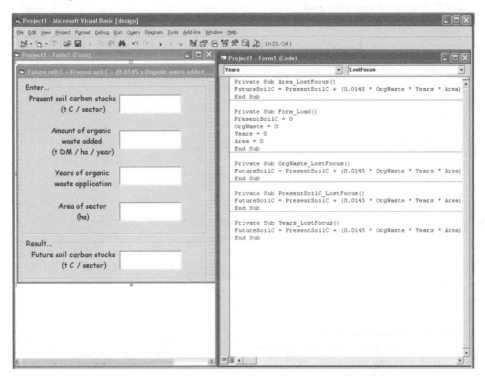

c. **If you can program in Fortran**, construct the complete model of the carbon stocks of Martian land with organic waste application in Fortran.

The program is given in Compaq Visual Fortran at Web link 2.22. The Fortran code is shown below.

```
C============================================================
C Simple Model: Future Soil C = Present Soil C +
C      (0.0145 x Amount of organic waste x Number of years)
C============================================================
C SECTCIN = Present Soil C in arable land (t C / sector)
C SECTAREA = Arable area of sector (ha)
C SOILCIN = Present Soil C (t C / ha)
C SOILCOUT = Future Soil C after given number of years (t C / ha)
C SECTCOUT = Future Soil C in arable land (t C / sector)
C WASTEADD = Amount of manure added each year (t DM/ha / year)
C NYEARS = Number of years of this manure management
      REAL SECTCIN
      REAL SECTAREA
```

```
      REAL SOILCIN
      REAL SOILCOUT
      REAL WASTEADD
      INTEGER NYEARS
C Get input variables from user
      PRINT*,'Enter present soil carbon stock (t C / sector):'
      READ(*,*)SECTCIN
      PRINT*,''
      PRINT*,'Enter arable area in the sector (ha):'
      READ(*,*)SECTAREA
      PRINT*,''
      PRINT*,'Enter amount of manure added per year (t DM / ha):'
      READ(*,*)WASTEADD
      PRINT*,''
      PRINT*,'For how many years does this management continue?:'
      READ(*,*)NYEARS
      PRINT*,''
C Calculate Present Soil carbon stock in t C / ha
      SOILCIN=SECTCIN / SECTAREA
C Calculate outputs from inputs
      SOILCOUT=SOILCIN+(0.0145*WASTEADD*NYEARS)
      SECTCOUT=SOILCOUT*SECTAREA
C Send output variable back to the user

      PRINT*,'============================================'
      PRINT*,'Future Soil C = ',SOILCOUT,' t C / ha'
      PRINT*,' = ',SECTCOUT,' t C / sector'

       PRINT*,'============================================'
      PRINT*,'... Press any key to continue'
      READ(*,*)
      END
```

    d. **If you can program in C++**, construct the complete model of the carbon stocks of Martian land with organic waste application in Visual C++.

The program is given in Microsoft Visual C++ at Web link 2.23. The C++ code is shown below.

```
//=========================================================================
// Problem2.3(d)
// Future soil carbon stock under organic waste application (t C / ha)
//  = Present soil carbon stock(t C / ha) +
//  (0.0145 x Amount of manure (t C t / DM / y) x Years of manure management)
//=========================================================================
#include "stdafx.h"

int main(int argc, char* argv[])
–
// Declare variables
float fSectorCInput;        //Present Soil C in arable land (t C / sector)
float fSectorArea;          //Arable area of sector (ha)
float fSoilCInput;          //Present Soil C (t C / ha)
float fSoilCOutput;         //Future Soil C after given number of years (t C / ha)
float fSectorCOutput;//Future Soil C in arable land (t C / sector)
float fWasteAdd;            //Amount of organic waste added each year (t DM / ha / year)
```

```
int nYears;              // Number of years of organic waste management
// Get input variables from user
printf("Enter present soil carbon stock (t C / sector): ");
scanf("%f",&fSectorCInput);
printf("\nEnter arable area in the sector (ha): ");
scanf("%f",&fSectorArea);
printf("\nEnter amount of manure added per year (t DM / ha): ");
scanf("%f",&fWasteAdd);
printf("\nFor how many years does this management continue? ");
scanf("%d",&nYears);
// Calculate Present Soil carbon stock in t C / ha
fSoilCInput=fSectorCInput / fSectorArea;
// Calculate outputs from inputs
fSoilCOutput=fSoilCInput + (0.0145*fWasteAdd*nYears);
fSectorCOutput=fSoilCOutput*fSectorArea;
// Send output variable back to the user
printf("==================================================\n");
printf("Future Soil C = %f t C / ha\n",fSoilCOutput);
printf(" = %f t C / sector\n",fSectorCOutput);
printf("==================================================\n");
return 0;
}
```

1.3 Solutions to problems in Chapter 3

3.1. **Evaluate the performance of a model** that simulates the optimum nitrogen fertiliser application rate to a range of crops. The following table gives the measured optimum nitrogen fertiliser application rate for a range of different crops. The rate recommended by the model is also given in the table.

Crop	Yield t ha^{-1}	Simulated Value	Measured value					
			Replicate 1	Replicate 2	Replicate 3	Replicate 4	Replicate 5	Replicate 6
Spring	5.1	120	138	130	125	115	112	110
Wheat	5.7	120	110	100	114	125	135	117
	6	120	100	105	102	108	110	130
	5.4	100	118	105	95	115	112	85
	5.8	100	115	97	133	122	145	116
	6	100	92	111	91	99	85	122
	5.8	140	145	157	155	132	138	121
	6	120	132	103	131	142	106	121
	5.4	100	87	102	97	112	118	105
	7.1	160	143	165	175	154	152	176
	7.4	150	157	132	165	175	132	151
	7.2	140	132	154	126	121	162	157
	6.5	160	176	165	160	145	147	141
	6.8	140	138	125	129	147	157	165

(continued overleaf)

(continued)

Crop	Yield t ha^{-1}	Simulated Value	Measured value					
			Replicate 1	Replicate 2	Replicate 3	Replicate 4	Replicate 5	Replicate 6
Winter	6	160	155	150	162	165	167	152
Wheat	6.5	159	150	145	157	167	170	175
	7.2	161	160	167	172	189	199	170
	7.4	181	165	200	173	171	168	184
	6.5	141	170	165	175	187	170	175
	7.2	139	170	167	172	152	155	170
	6.7	162	166	155	145	182	187	177
	6.4	141	154	132	161	123	153	125
	7	161	175	153	145	154	180	174
	7.8	181	164	167	178	197	193	178
	8.5	200	186	182	194	214	218	201
	8	180	174	194	204	176	164	177
	7.4	160	162	145	178	153	169	180
	7.2	180	200	202	187	193	197	173
Winter	6	160	155	150	162	165	167	152
Oilseed	3.5	220	175	165	186	155	180	185
Rape	4	250	200	195	193	205	169	172
	4.2	263	205	210	190	203	202	185
	3.9	250	188	182	177	155	201	200
	4.1	220	181	175	172	203	172	155
	4.2	220	162	125	143	135	133	111
	3.5	220	154	142	135	154	137	175
	3.6	220	176	159	154	171	145	197
	3.2	200	165	127	138	143	176	146
	4	241	201	197	176	186	180	173
	4.7	279	220	243	237	222	254	265
	4.4	260	210	212	254	220	223	201
	2.5	141	121	100	123	75	89	92
	3.2	179	134	156	123	111	142	152

Plot the results. What do the plots show? Determine the coincidence and association between the simulated and the measured values. Is the model providing a good estimate of the optimum nitrogen application rate? Discuss how well the model performs overall, and for each individual crop. The MODEVAL spreadsheet (in Microsoft Excel, see Web link 3.1) can be used (if required) to calculate the statistics.

Download MODEVAL from the website at Web link 3.1 Type the data into the MODEVAL spreadsheet. Plot the mean replicated measurement against the simulated value using the 'Chart Wizard'.

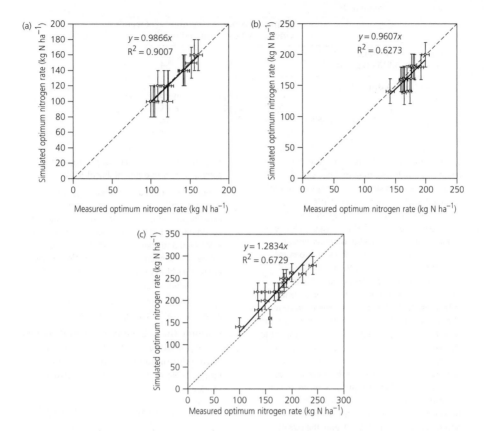

The mean measured nitrogen rate against the simulated nitrogen rate for (a) spring wheat, (b) winter wheat and (c) winter oilseed rape.

The plots of the measured against the simulated values of the optimum nitrogen rate of spring wheat against the mean measured values show good coincidence and association in the values. The simulated values for winter wheat also show high association, but the coincidence is reduced. The values simulated for winter oilseed rape show lower coincidence and association than either spring wheat or winter wheat.

	Spring wheat	Winter wheat	Winter oilseed rape	Overall
Coincidence				
Error excluding				
measurement error				
LOFIT	4022	13 089	260 847	277 958
F	0.10	0.39	4.74	0.73
F (critical value 5%)	1.87	1.87	1.83	1.44
Total error				
RMSE	5%	7%	31%	21%
$RMSE_{95}$	12%	9%	10%	10%
Bias				
E	1	4	−30	−9
E_{95}	12	8	10	10

(continued overleaf)

(*continued*)

	Spring wheat	Winter wheat	Winter oilseed rape	Overall
Average error				
RMS (kg N ha^{-1})	7	12	54	33
Association				
r	0.95	0.79	0.87	0.80
t	10.47	4.49	6.45	8.62
t (critical value 5%)	2.18	2.18	2.16	2.02

The F-test on the LOFIT indicates that the coincidence between the simulated and the measured values is significant for spring wheat and winter wheat, but not for winter oilseed rape. Overall the LOFIT is not significant. This is reflected in the value of RMSE, which is less than RMSE$_{95}$ for spring wheat and winter wheat. However, overall the value of RMSE is greater than RMSE$_{95}$. This reflects the importance of removing the error due to the variation in experimental measurement from the statistic, as in this case RMSE has given a misleading result. The simulated values of spring wheat and winter wheat are not significantly biased with respect to the measured values, indicating that the model includes processes in the correct proportions. For winter oilseed rape, the bias is significant and negative. This indicates that some processes of nitrogen dynamics in winter oilseed rape significantly underestimate the amount of nitrogen in the soil profile. Overall the bias is insignificant. The average errors associated with the simulations are low for spring wheat and winter wheat (7 and 12 kg N ha^{-1}, respectively), whereas the average error for winter oilseed rape is very high, namely, 54 kg N ha^{-1}, pulling the overall average error up to 33 kg N ha^{-1}. The simulations of all crops show significant association between the simulated and measured values (as indicated by the value of the Student's

t being less than the value of the Student's t corresponding to a probability of 5%).

Overall, the model does not provide the simulation of measured values to within experimental error. However, the simulations for spring wheat and winter wheat are significantly associated and coincident with the measured values. The simulations for winter oilseed rape are neither highly coincident nor associated with the measured values.

Spreadsheets in Microsoft Excel showing the above calculations are given at Web link 3.5.

3.2 **Examine the behaviour of a simple model** of the growth of a rabbit population in a warren during summer. The number of rabbits is estimated using the following model:

$$R = R_{start} + \exp(k_1 \times t) + \exp(k_2 \times t),$$

where R_{start} is the measured number of rabbits (500 rabbits), t is the time in weeks since the number of rabbits was measured, and k_1 and k_2 are parameters describing the growth in the population ($k_1 = 0.5$ and $k_2 = 0.02$).

Plot the calculated change in the rabbit population with time from zero to ten weeks.

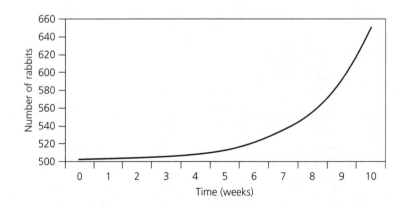

Use a grid search to investigate the sensitivity of the model to changes in R_{start}, k_1 and k_2 within the range of $\pm 20\%$ with step sizes of 5%. Plot the results and calculate the sensitivity index for each combination of R_{start}, k_1, and k_2. Plotted at 10 days:

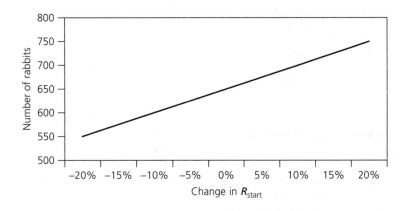

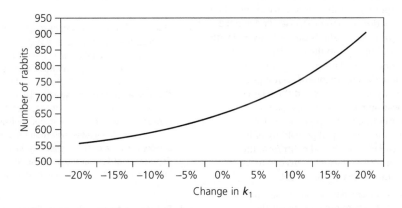

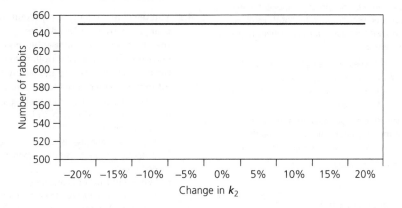

Use a one-at-a-time analysis to calculate the sensitivity index for R_{start}, k_1, and k_2.

Input	Sensitivity index
R_{start}	0.27
k_1	0.39
k_2	0.0002

Can the model be simplified in any way?

The exponential expression including k_2 could be omitted.

A spreadsheet in Microsoft Excel to complete the above calculations is given at Web link 3.6.

3.3 **Determine the importance of the model components** to the uncertainty of the rabbit population growth model (see Problem 3.2). The number of rabbits can be measured with 20% accuracy, and the population growth parameters are known to be 50% accurate. Define the importance of each input using the importance index and the relative deviation ratio, assuming a uniform distribution for the probability density functions (i.e., equal probability of finding a parameter at each grid point from the mean). What research is needed to reduce the uncertainty of the model?

As the uniform distribution of the probability density functions can be assumed, the importance of the model components can be calculated very simply in Microsoft Excel. A spreadsheet showing the calculations is given at Web link 3.7.

Obtain a series of nine input values at -1, -0.75, -0.5, -0.25, 0, 0.25, 0.5, 0.75 and 1 times the accuracy of the selected measurement (R_{start}. k_1 and k_2).

Calculate the input value for the measurement and the resultant output value using the equation $R = R_{start} + \exp(k_1 \times t) + \exp(k_2 \times t)$.

Calculate the importance index from

$$\text{Importance index} = \frac{\sum_{i=1}^{n} (I_i - \bar{I})^2}{\sum_{i=1}^{n} (P_i - \bar{P})^2},$$

and the relative deviation ratio from

Relative deviation ratio

$$\frac{\bar{O} \times \sqrt{\sum_{i=1}^{n} (P_i - \bar{P})^2}}{\bar{P} \times \sqrt{\sum_{i=1}^{n} (O_i - \bar{O})^2}}.$$

This gives the following results.

	R_{start}	k_1	k_2
Importance index	1.00	6.46×10^{-8}	0.01
Reciprocal of the importance index	1.00	1.5×10^7	150
Relative deviation ratio	0.77	3.03	4×10^{-4}

The importance index (and its reciprocal) suggests that the k_1 induces a larger change in the output per unit change in the input value than either R_{start} or k_2, and R_{start} appears to be the least important input. However, this index takes no account of the average size of the inputs and outputs, and so is only useful in comparing the importance of inputs of similar magnitudes. The relative deviation ratio provides a more accurate picture, reflecting the greater importance of k_1 over the other inputs, as well as indicating that R_{start} is more important than k_2.

1.4 Solutions to problems in Chapter 4

4.1. **You develop a scientific representation** of goat productivity in East African smallholdings. You will use this model to determine the factors controlling longevity, the production of young and lactation. This information will be used to prepare a scientific paper on the controls of goat longevity and productivity in East Africa.
   a. Who is the end-user?
      The primary end-user is you, the model developer. You are also the secondary end-user.
   b. What will the application do?
      The application will run the model for a range of factors. The numbers produced will allow you to determine the factors that control longevity and productivity.

c. How should the application guard against input error?

You are the primary end-user, so protection against input error can be limited to warning and error messages produced from within the model.

d. How can the application guard against misinterpretation of the results?

You are the primary end-user, so protection against misinterpretation of the results is unnecessary. You will analyse the output numbers, before further interpretation.

e. What documentation is needed?

The documentation should include the following: a clear statement of the model objectives; a description of the structure, hypotheses, assumptions and boundary conditions underlying the model; a report of the mathematical formulae used; a fully-commented computer implementation; and an explanation of how the model is run.

4.2. You develop an expert system to provide nitrogen fertiliser recommendations on UK farms.

a. Who is the end-user?

The primary end-user is the farmer or advisor who will use the model themselves.

b. What will the model do?

The application will run the model for the inputs given by the farmer or advisor. The inputs might include the soil type, the average yearly or spring rainfall, the crop to be grown, and how much organic fertiliser is added. The model will then provide the farmer or advisor with a suggestion of how much fertiliser to apply and when.

c. How can the application guard against input error?

The application will have a graphical user interface that allows the farmer to enter specific inputs. It will prompt the farmer to specify the units he/she is using and it will give the usual ranges for the input values. It might disallow values outside this range and give the user warnings. There will be a help system to allow the farmer to get immediate help and a user manual to read away from the computer. A tutorial would also be very useful.

d. How can the application guard against misinterpretation of the results?

The application will give simple expert advice that will not allow scope for the user to misinterpret the results, and will carefully specify the units used. The model will then provide the farmer or advisor with a suggestion of how much fertiliser to apply and when.

e. What documentation is needed?

The graphical user interface will have built-in help and will also access, and cross-reference to, the user manual. Technical documentation will also be available, but will not be necessary to use the application.

4.3. You develop a decision support system to provide pesticide application advice for use on tropical palm plantations.

a. Who is the end-user?

The primary end-user is the plantation manager, who will use the model themselves.

b. What will the model do?

The application will run the model for the inputs given by the plantation manager. The inputs might include the soil type, information about the temperature and rainfall, the age of the palms in the plantation, fertilisation and other management. The model will then provide the plantation manager with a range of options for pesticide application (which pesticides and when to apply them), based on predictions of pest outbreak likelihood and the efficacy of the pesticides.

c. How can the application guard against input error?

As for the expert system suggested in Problem 4.2, the application will have a graphical user interface that allows the user to enter specific inputs. It will prompt the plantation manager to specify the units he/she is using and it will give the usual ranges for the input values. It might disallow values outside this range and give the user warnings. There will be a help system to allow the user to get immediate help and a user manual to read away from the computer. A tutorial would also be very useful.

d. How can the application guard against misinterpretation of the results?

The application will give recommendations. Unlike the expert system, this will not be a single prescriptive statement. Instead, it might give the likelihood of pest outbreak, the likely impact of the pesticide if applied on a number of specified dates and the likelihood that the pesticide remains in the environment long enough to be effective (e.g., it might be washed away in the middle of the rainy season). The system might also assess non-target hazard, to allow the manager to assess the risk to other wildlife.

e. What documentation is needed?

As for the expert system described in Problem 4.2, the graphical user interface will have built-in help and will also access, and cross-reference to, the user manual. Technical documentation will also be available, but will not be necessary to use the application.

4.4. You wish to develop a risk assessment scheme for assessing the impacts of a new herbicide to get rid of weeds from Martian fern fields on Mars.

a. Who is the end-user?

The primary end-user will be you, the scientist who advises the Martian Herbicide Regulator.

b. What will the model do?

The application will contain many models. In order to perform the overall risk assessment, the system will need to assess the probability of the concentration in the environment exceeding the environmentally safe concentration.

c. What types of data might you need to set up the risk assessment scheme?

You will need to know the toxicity of the herbicide to Martian fern (the crop species—hopefully this is low or it will make a very poor herbicide!), and to non-target organisms, including other plants and animals that might be exposed to the herbicide. This will be needed to determine the environmentally safe concentrations. You also need to know at what concentration the active ingredient is applied and how it is applied to the field (spray, solid pellets, etc.). You will also need physical or chemical data on the compound, for example, how soluble or volatile it is. This will be necessary to determine the likely fate and concentrations of the active ingredient in the water, air and soil of the Martian environment. The company producing the herbicide might need to provide you with this information, as they do for the regulators on Earth.

d. What types of model might you use as part of the risk assessment scheme?

For assessing toxicity to the crop plant and non-target organisms, you will use data, if available, and if there are similar compounds already licensed you might use quantitative structure activity relationships (QSARs) to estimate the toxicity. You will apply safety factors (in themselves, very simple models) to derive environmentally safe concentrations of the herbicide, perhaps measured by a predicted no-effect concentration (PNEC). For assessing the fate and distribution of the compound, you will need a model to predict the dispersal in the environment from the point of application and models to determine the part of the environment where the herbicide will most likely end up (e.g., the soil, atmosphere or water). The models will allow the concentration in various parts of the environment to be estimated. These models may be run many times with different inputs (e.g., different weather patterns) to determine the probability of the environmentally safe concentration being exceeded.

e. How can the application guard against input error?

You are the primary end-user, so protection against input error can be limited to warning and error messages produced from within the model.

f. How can the application guard against misinterpretation of the results?

You are the primary end-user, so protection against misinterpretation of the results is unnecessary. You will analyse the output numbers, before further interpretation.

g. What documentation is needed?

The documentation should include the following: a clear statement of the model objectives; a description of the structure, hypotheses, assumptions and boundary conditions underlying the model; a report of the mathematical formulae used; a fully-commented computer implementation; and an explanation of how the model is run. If you provide a Web-based tool for the public to use (to increase transparency and build trust with the Martian public), then the graphical user interface will have built-in help and will also access, and cross-reference to, the user manual. Technical documentation will also be available, but will not be necessary to use the application. All of the background data accessed by the Web-based tool should be accessible to the user.

4.5. How would you develop and use a spatially-explicit application of a model to estimate the effects of climate change on forest growth in the northern hemisphere?

a. Who is the end-user?

The primary end-user is you, the model developer. You are also the secondary end-user.

b. What will the model do?

The model will estimate the impact of

changing temperatures and changing patterns of precipitation and evapo-transpiration on tree growth in the northern hemisphere. Since different tree types react differently to temperature and water balance, a number of tree plant functional types will be needed. The simplest model might divide the northern hemisphere up into a regular grid and use satellite images of forest inventory data to assign the proportion of each grid cell occupied by each tree functional type. The model will then use climate data (from climate models) to assess how well trees grow under these future conditions. More complex models might allow certain tree types to become more common than others as the climate changes (i.e., to allow competition between tree functional types). Yet more complex models might allow individual trees to compete using individual-based models. Other models might include the changing risk of fire, tree death and pest attack in the projections. What is common to all approaches is that models are run many times, once for each area of land, using a spatial database of input variables to drive the models.

c. How can the application guard against input error?

Input error is inherently more likely for spatial applications of models, for a number of reasons.

- Due to the large volume of data, errors are inherently more difficult to spot.
- At least some of the spatial datasets come from secondary sources, that is, data that you did not collect yourself. This means that
  i. you have no control over data quality,
  ii. the format may be confusing (e.g., many datasets multiply values with one decimal place in the data by 10 to remove the decimal place—i.e., a mean monthly temperature of 1.5 °C will be stored as 15 in the database and thereby reduce the size of the file),
  iii. the data may be derived from model outputs and the uncertainty

associated with the values may be unknown, and
  iv. the units may be different from those used by your model.

- There are likely to be missing values for some input parameters in some cells within the grid. The missing value code may differ among datasets.
- There are a variety of geographical projections used, and there are different conventions as to where the grid is measured from (e.g., the centre of the grid cell or the bottom left corner of the grid cell).
- Data for different spatial parameters are often available at different spatial scales, and using different spatial formats (e.g., some are in grid format and some in polygons).

To guard against input error, data must first be converted to a compatible projection and the grids cells or polygons overlaid. This is most easily accomplished within a geographical information system. Plotting the values on a map allows outliers to be spotted. Careful checks on the units used and on any multiplication factors need to be made. Where possible, the uncertainty in the input data should be quantified. This is essential when interpreting the results. All missing values need to be converted to a standard code that is recognised by the model.

d. How can the application guard against misinterpretation of the results?

You are the primary end-user, so protection against misinterpretation of the results is unnecessary. You will analyse the output numbers, before further interpretation.

e. What documentation is needed?

The documentation should include the following: a clear statement of the model objectives; a description of the structure, hypotheses, assumptions and boundary conditions underlying the model; a report of the mathematical formulae used; a fully-commented computer implementation; and an explanation of how the model is run.

■ GLOSSARY

Algorithm A procedure (a finite set of well-defined instructions) in a computer program for accomplishing some task which, given an initial state, will terminate in a defined end state.

Alternative hypothesis A possible alternative to the null hypothesis.

Application of a model 1. The definition of a model by its use—functional or mechanistic. 2. The process of constructing a model into a working tool.

Association The quantification of how well two sets of values follow the same trend.

Assumption A proposition that is taken for granted; in other words, one that is treated for the sake of a given discussion as if it were known to be true.

Bayesian statistics Statistics that incorporate prior knowledge and accumulated experience into probability calculations, based, for example, on the previous year's observations.

Bias The tendency of the simulated values to overestimate or underestimate the corresponding measurements.

Boundary conditions The range of different conditions within which the model can be expected to simulate the result to within the required level of accuracy.

Cellular automata The separation of continuous space into discrete cells, which react, by a series of rules or relationships, to the local conditions around the cell (e.g., the condition of neighbouring cells).

Coincidence The quantification of how well two sets of values match each other.

Compiled code The translation of text written in a computer language (the source language) into another computer language (the target language). The original sequence is usually called the source code and the output is called object code. Commonly, the output has a form suitable for processing by other programs (e.g., a linker), but it may be a human-readable text file.

Computer memory The place where the computer holds the current programs and data that are in use.

Computer program A list of instructions that explicitly implement an algorithm; a collection of source code and libraries which have been compiled into an executable file or otherwise interpreted to 'run' in (active) computer memory, where it can perform both automatic and interactive tasks with data.

Conceptual model The list of hypotheses, assumptions and boundary conditions that define the model.

Confidence interval The estimated range of values that is likely to include the value to be estimated.

Controls The factors controlling the progress of a process.

Decision support system A computer program that provides the user with information that supports the decisions they need to make.

Default values The values that a model will use in the absence of alternative input values being entered by the user.

Descriptive model A model that describes observations within the conditions of the current experiment.

Deterministic model A model producing an output variable that is a single value (as opposed to a range of values).

Development of a model The process of constructing a model from an idea into a conceptual model, through a mathematical model, and finally into a computer model.

Differential uncertainty analysis An uncertainty analysis involving the partial differentiation of the model in its aggregated form.

Differentiation This expresses the rate at which a variable changes with respect to the change in another variable on which it has a functional relationship.

Discontinuities These occur when small changes in the inputs to a model or function result in usually large and stepped changes in the output.

Documentation Any communicable material (such as text, video, audio, etc.) that is used to explain some attributes of the model or system.

Dynamic model A model in which variables change during the model run.

End-user The person who uses the model or system.

Error bars A depiction of the uncertainty around a value plotted on a graph.

Evaluation of a model The process of testing the model performance.

Executable file A computer file, the contents of which are interpreted as a program by a computer.

Expert system A computer program that will give direct advice to the user.

Factorial sensitivity analysis A sensitivity analysis in which more than one parameter is adjusted at a time.

Feedbacks The output factors of a process that are passed (fed back) to the input. Negative feedback reverses the direction of change, so stabilising the process. Positive feedback increases the change, so causing the output to change even more in the same direction.

Fixed parameters The numbers in a model that stay constant over different conditions.

Formatting The order and form of entered data, that is, the number of spaces, dividing characters and the type of variable being entered.

Functional model A model that represents observations without a description of the underlying mechanism.

Geostatistical model A statistical model that can be used to describe the spatial distribution of the results.

Global optimum A selection from a given domain that yields either the highest value or the lowest value (depending on the objective), when a specific function is applied.

Goodness-of-fit This describes how well the modelled values fit a set of observations.

Graphical analysis The graphical depiction of the measured and simulated values to allow a qualitative comparison.

Hardware The physical components of a computer system.

Help system A component of the modelling system that provides information to help the user to run the model and interpret the results.

Hypothesis A suggested explanation of a phenomenon.

Independent measurements The measured values that were not used to develop the model.

Input variables The numbers that change with each model run.

Inputs The information needed to drive the model.

Integration The process of finding the area under the curve that describes the relationship between two variables.

Iteration The repetition of a process within a computer program.

Library A collection of subprograms used to develop software.

Mathematical model A simplified representation of reality described using mathematical formulae.

Mechanistic model This represents observations with a description of the underlying mechanism.

Model A simplified representation of reality.

Monte Carlo sampling Random sampling from within the potential distribution of the input values.

Neural network A piece of software that is 'trained' by presenting it with examples of the input and the corresponding desired output. Neural networks mimic the vertebrate central nervous systems, to develop rules about relationships between inputs and outputs.

Null hypothesis A negative statement of the thing you want to test.

Ockham's razor Of the two competing explanations, both of which are consistent with the observed facts, we regard it as right and obligatory to prefer the simpler (Barker, 1961, p. 273, listed in the 'References' section of Chapter 2).

One-at-a-time sensitivity analysis A sensitivity analysis in which the inputs are varied one by one.

One-at-a-time uncertainty analysis An uncertainty analysis in which the inputs are varied one by one.

Outputs The information produced by the model.

P value The probability that the null hypothesis is true.

Parameter optimisation A method to obtain the model parameters by adjusting the values of the selected parameters so that the difference between the simulated and measured outputs is minimised.

Parameters See **Fixed parameters**.

Predictive model A model that can extrapolate beyond the scope of the experiment, and provides results that extend beyond the current observations.

Principle of parsimony Of the two competing explanations, both of which are consistent with the observed facts, we regard it as right and obligatory to prefer the simpler (Barker, 1961, p. 273, listed in the 'References' section of Chapter 2).

Probabilistic algorithm An algorithm that employs a degree of randomness to provide simulations of the 'average case'.

Probability density function A function that gives the distribution of the possible values of an input, and so can be used to randomly select the input value from within this distribution.

Process-based model A model that explicitly contains information about the processes in the system.

Propagation of errors When errors in the outputs from one unit of a model act as errors in the inputs to another unit of the model.

Qualitative value A value that describes the nature of the output.

Quantitative analysis This determines the accuracy of a simulation, that is, how well the simulated values match the measured data.

Quantitative value A value that provides a numerical measurement or count.

Regional-scale modelling A spatial application where the model simulates a spatial region.

Risk assessment The use of a model to assess the risks associated with a particular course of action.

Scientific representation The use of a model to interpret the experimental observations of a system and improve our understanding of those observations.

Scope of a model The use of a model to provide results within or outside the experiment used to develop it; the scope is named descriptive or predictive, respectively.

Sensitivity analysis An evaluation of the behaviour of a model, used to determine whether the model responds in the expected way to changes in the conditions of the simulation.

Simulated annealing A generic probabilistic algorithm to locate a good approximation to the global optimum of a given function in a large search space.

Software The programs that enable a computer to perform a specific task.

Spatial model A model that is applied to represent differences in physical space.

Source code Any series of statements written in some human-readable computer programming language.

Spreadsheet program A rectangular table (or grid) of information that allows the calculation of new values from the entered data.

Stakeholder A person or organisation that has a legitimate interest in a project.

Static model A model in which the variables are fixed for any model run.

Statistical model A model that derives the mathematical equations used directly from the measured data using standard statistical techniques.

Stochastic model A model in which the output variables are expressed as a range of possible values.

Total difference The average difference between the simulated and measured values.

Uncertainty analysis A method to resolve the importance of the model components.

User interface The means by which people (the users) interact with the model.

■ INDEX

A

acceptable error 75, 87–91
application:
    functional 11, 13
    mechanistic 11, 13
association, analysis of 70–71,
    81, 94–98
assumptions 33
axes 75

B

Bayesian:
    model 17
    statistics 39
bias 81, 91–93
boundary conditions 33–34
bugs 56, 57

C

cellular automata 17, 40
coefficient of determination
    94
coefficient of residual mass 94
coincidence, analysis of 70, 81,
    84–95
compilation of computer code
    55
complexity of models 29
computer model 45–60
conceptual model 22, 30, 31–35
confidence interval 74, 87–91
controls on processes 123
correlation coefficient,
    sample 94–96, 108

D

decision support systems 120,
    128–134, 148
descriptive model 11, 12
deterministic model 11, 12
discontinuities in processes 125
differential analysis 110
documentation 120, 127,
    132–133, 138, 144–145

E

end-user 120, 123–124, 129–130,
    135–136, 139–140
    primary 123, 129, 136, 139
    secondary 123, 129, 136, 139
error messages 121
errors in format 57
evolutionary algorithms 40
executable file 55
expert system 120, 128–134

F

F test 93–94, 96
feedbacks on processes 125
fixed parameters 36, 49
functional model 11, 13

G

geographical information
    systems 120, 142
geostatistical model 16
goodness-of-fit 40
graphical analysis 70, 71–80
graphical user interface 121
grid search 107

H

high-level programming
    languages
    advantages in modelling 59
    disadvantages in modelling 59
    use in modelling 55–59
hypothesis:
    alternative 10
    list 33
    null 10

I

importance index 111
independent data 83
input error 120, 126, 130, 137,
    143–144
input variables 36, 49
inputs:
    dynamic 11, 12
    static 11, 12

L

Latin-hypercube 111
linear scale 75
linking object files 55
lack of fit statistic 93–94
logarithmic scale 75
logical analysis 106
look-up tables 136

M

machine code 55
mathematical approach:
    Bayesian 17, 39
    choice of 38–45
    cellular automata 17
    geostatistical 16
    neural network 17
    statistical 16
mathematical equations 46
mathematical model 22, 35–40
mechanistic model 11, 13
mean difference 92–93
mean measurement 88
misinterpretation of results 120,
    126–127
model:
    classification of 10–14
    complexity 29
    computer 45–60
    conceptual 22, 30, 31–35
    definition of 2–3
    descriptive 11, 12
    deterministic 11, 12
    documentation 120, 121
    functional 11, 13
    geostatistical 16
    mathematical 22, 35–40
    mechanistic 11, 13
    predictive 11, 12
    process-based 18
    qualitative 11
    quantitative 11, 12
    scope of 5
    sensitivity 108

model: (*cont.*)
  sharing models 28
  spatially explicit 120
  static 11, 12
  statistical 16
  stochastic 11, 12, 74
  working 22
modelling efficiency 94
Monte Carlo techniques 79,
  111

N

negative correlation 96
neural network 17, 38–39

O

object files 55
Ochham's razor 29–30
outliers 73
outputs:
  deterministic 11, 12
  qualitative 11
  quantitative 11, 12

P

P value 90
parameters, fixed 36
parameter optimisation 39
parsimony, principal of 29–30
partial correlation coefficient
  112
plots:
  of accuracy of
    simulation 72–76
  of behaviour of model
    components 76–78
  of importance of model
    components 78–80
Plackett-Burnham techniques
  111
PNEC 134
polar charts 79
positive correlation 96
predicted no effect
  concentration 134
probability density function 107,
  110
programming languages:
  Basic 56
  C++ 56, 57

Fortran 56, 57, 59
Java 56
Visual Basic 56, 57, 58,
  59
Visual Fortran 58
Visual C++ 57, 58, 59
propagation of errors 29
predictive model 11, 12
process-based model 18

Q

QSARs 134
qualitative model 11
quantitative analysis 70, 81–105
quantitative model 11,12
quantitative sensitivity
  analysis 106
quantitative structural activity
  relationships 134

R

rank transformations 112
real life applications 147–150
regression analysis 43, 108, 111,
  112
relative deviation method 111
relative error 92
reverser modelling 125
risk 17
risk assessment 120, 134–138
root mean squared deviation
  86
root mean squared error 86–87

S

sample correlation
  coefficient 94–96, 108
scope:
  descriptive 11, 12
  predictive 11, 12
scientific representation , use of a
  model for 120, 122–128
sensitivity analysis 71, 106–109
  factorial 107
  global 107
  local 107
  multiple 107
  one-at-a-time 107
  regression 108
  subjective 106

sensitivity:
  coefficient 108
  index: 108
significance:
  of bias 92
  of sample correlation
    coefficient 96
  of total error 87–91
simulated annealing 40
source code 55–56
spatially explicit
    applications 120, 139–146
specialist modelling software
  advantages in modelling 53
  disadvantages in modelling 55
  use in modelling 52–55
spreadsheets:
  advantages in modelling 49, 52
  disadvantages in modelling 49,
    52
  use in modelling 46–52
standard error 74, 88
state variables 46
static model 11, 12
statistical fitting 38
statistical model 16
stochastic model 11, 12
Student's *t* 88–89, 93
subjective sensitivity analysis 106
systematic error 73, 82

T

total difference 81, 84–87
total error 81, 84–87
trends 73, 81, 96

U

uncertainty analysis 71, 106,
    109–113
  local 110
  one-at-a-time 110
  partial differentiation 110
  regression 111

V

variables, inputs 36

W

warning messages 121